AUGSBURG COLLEGE AND SEMINARY
LIBRARY - MINNEAPOLIS 4, MINN.

Date Loaned

THE MEDITERRANEAN

By the same author

POST-WAR BRITAIN
AMERICA COMES OF AGE
ENGLAND'S CRISIS
IMPRESSIONS OF SOUTH AMERICA
EUROPE'S CRISIS
CANADA
SUEZ AND PANAMA

THE MEDITERRANEAN

by

ANDRÉ SIEGFRIED

Translated from the French by
DORIS HEMMING

DUELL, SLOAN AND PEARCE
NEW YORK

ALL RIGHTS RESERVED, INCLUDING THE
RIGHT TO REPRODUCE THIS BOOK OR
PORTIONS THEREOF IN ANY FORM

PRINTED IN GREAT BRITAIN

CONTENTS

	TRANSLATOR'S FOREWORD	9
	PREFACE	11
	INTRODUCTION	25
I	GEOLOGICAL ORIGINS	37
II	THE MEDITERRANEAN AS A SEA	49
III	THE MEDITERRANEAN CLIMATE	57
IV	THE FLORA OF THE MEDITERRANEAN	67
V	LANDSCAPE AND SCENERY	77
VI	THE MEDITERRANEAN RACE	91
VII	TRADITIONAL AGRICULTURAL METHODS	107
VIII	INDUSTRIALISED CULTIVATION	118
IX	PROPERTY AND OWNERSHIP	128
X	MINERAL RESOURCES	138
XI	INDUSTRIAL DEVELOPMENT	153
XII	THE MEDITERRANEAN PORTS	164
XIII	TRADE CURRENTS	179
XIV	THE BALANCE OF TRADE	192
XV	THE SANITARY DEFENCE OF THE WEST	198
XVI	THE MEDITERRANEAN'S PLACE IN THE WORLD	211

ILLUSTRATIONS

Position of the Mediterranean on the map of the world	26
The latitude and longitude of the Mediterranean	28
The structure of the Old World	38
Ancient massifs in the Mediterranean area	40
The Hercynian chain	41
The Alpine folds and their Asiatic equivalents	43
The position and direction of the Alpine folds	45
Natural boundaries of the Mediterranean region	61
The geographical distribution of the European races according to von Eickstedt	93
The distribution of the races of the Old World according to Haddon	94
The three zones from which people of the present day originated according to the *French Encyclopaedia*	100
Regions of intensive cultivation	110
The Mediterranean's share in the world production of various minerals	139
Coal production in the Mediterranean	141
Iron and chromium ore production in the Mediterranean	141
Lead and zinc production in the Mediterranean	143
Copper production in the Mediterranean	143
Mercury production in the Mediterranean	145
Production of phosphates and potash in the Mediterranean	145
Sulphur production in the Mediterranean	147
Bauxite production in the Mediterranean	147
Petrol production in the Mediterranean	149
Industrial centres in the Mediterranean region	161
The harbours of Piraeus in ancient times, according to Hérubel	165
The ancient harbour of Carthage, according to Hérubel	166

ILLUSTRATIONS

The port of Genoa in 1875	167
The port of Genoa in 1890	167
The port of Genoa in 1929	169
The port of Marseilles	170
The port of Caronte	171
The port district of Marseilles	171
The port of Sète	173
Complementary maritime routes (coal against heavy industrial products)	181
Oil trade routes through the Mediterranean	183
Winter resort areas according to Maximilien Sorre	201
Endemic malarial districts, according to Maximilien Sorre	202
Sources from which diseases are introduced to the Mediterranean area, according to Maximilien Sorre	206

TRANSLATOR'S FOREWORD

IT is with pleasure, indeed with relief and thanksgiving, that we present another of Professor André Siegfried's books to his English-speaking public. In these pages devoted to many different aspects of his beloved Mediterranean he is penetrating and keen, observant and enthusiastic, and still abounding with energy.

And yet the German occupation must have affected him. A passage from a letter received from him shortly after the liberation of Paris reveals the tremendous strain on sensitive French people: 'At last we are able to breathe freely after four years of veritable oppression. I carried on my lectures throughout the war at both l'Ecole des Sciences Politiques and le Collège de France. The Germans left me alone, and so I was never asked to say or forbidden to say certain things — but in any case I should never have consented. Yet I always felt that spies were listening to every word I said, so I had to be extremely prudent. However, by saying what I had to say very quietly — *avec douceur* — I succeeded in expressing many ideas which certainly would never have passed the German censor. I have thus lived through the years of enemy occupation without, I hope, compromising my intellectual integrity in any way.'

A New York paper, describing conditions and personalities in Paris, singled out Professor Siegfried as a man noted for his moral courage. 'He held his head high, and kept telling his compatriots that these were days when all Frenchmen should stand shoulder to shoulder in their loyalty to the Republic.'

Recognition of his qualities soon followed the end of the occupation. In October 1944 he was elected a member of the French Academy by acclamation — the highest honour that France has to bestow upon her literary men.

In March 1945 he was invited by the Royal Institute of International Affairs to give a lecture at Chatham House in

TRANSLATOR'S FOREWORD

London. It was a happy occasion for reviving past friendships and cementing his ties with this group of British intellectuals.

In the summer of 1945 he was one of the delegates appointed to represent France at the United Nations Conference in San Francisco. He was obliged to leave before the end, as he had previously agreed to give the Romanes Lecture at Oxford, on June 5th.

While he was still in San Francisco the Canadian Government invited him to visit Canada that summer, and as an inducement offered to put everything at his disposal 'from the Alaska boundary to the Atlantic seaboard'. With amazing energy he immediately crossed the Atlantic again, accompanied by his wife, and together they made an exhaustive tour of Canada. His interviews and observations confirmed his earlier impressions and justified various predictions made in his book on that country which was published in 1937. We are hoping that he will produce a standard work on Canada, especially as the knowledge gleaned on this visit has already taken shape in his course of post-graduate lectures last winter at the Collège de France.

DORIS HEMMING

PREFACE

WE are now in 1947, and the Mediterranean still presents one of the chief political problems of the globe. Even after so many catastrophes, wars and upheavals, this highway is always of outstanding importance, for it is one of the maritime routes — one of the great airways — without which there can be no control of international trade. Once war is declared command of the Mediterranean becomes a vital necessity to each of the belligerents.

Renan, in welcoming Ferdinand de Lesseps to the French Academy in 1885, reminded him of the classic saying which must frequently have crossed his mind: 'I come not to bring peace but a sword.'

'The Bosphorus by itself has been enough to keep the whole civilised world embarrassed up to the present,' he went on, 'but now you have created a second and much more serious embarrassment. Not merely does the canal connect two inland seas, but it serves as a communicating passage to all the oceans of the globe. In case of a maritime war it will be of supreme importance, and everyone will be striving at top speed to occupy it. You have thus marked out a great battlefield for the future.'

These prophetic words are as true today as they were when they were spoken. In the war of 1939, just as in the war of 1914, the Isthmus of Suez was still the aim, the bomb target, one of the points which, as Renan predicted, the belligerents tried to obtain as quickly as possible, for the Mediterranean remained, even if they could not always make use of it, an area essential to victory on the continent.

This sea is unique, and in studying it one is immediately struck by the fact that it will always be at the heart of the world's political problems. Ever since there has been a Western civilisation distinct from the civilisations of Asia, the Mediterranean has occupied a central place in the history of the world. When-

PREFACE

ever it lost this central position, as it did after the discovery of the route around the Cape of Good Hope, it was only to return to it, as in 1869 when the Suez Canal was opened. When it seemed to lose it once more as the aeroplane and the submarine threatened its security, it regained its importance when peace was established.

After the Treaty of Versailles there was a certain school of thought in Great Britain which was inclined to abandon the Mediterranean in favour of the South African route, but events soon forced it to drop the idea. Even today, when India and Pakistan have become independent dominions, we can already foresee that a powerful England will never lose interest in this sea.

How well Conrad appreciated its naval appeal: 'The cradle of oversea traffic and of the art of naval combats, the Mediterranean, apart from the associations of adventure and glory, the common heritage of all mankind, makes a tender appeal to the seaman. It has sheltered the infancy of his craft. He looks upon it as a man may look at a vast nursery in an old, old mansion, where innumerable generations of his own people have learned to walk....'

So it has been from the beginning of the Western world. Now, however, when this world is heavy with the weight of centuries, already reaching maturity, perhaps even old age, it remains in the forefront of the news in the most astonishing way. If a third world war were to break out tomorrow we should find that control of the Mediterranean would again be essential to victory.

After Vasco da Gama had discovered a new way to the Indies in 1498, the Mediterranean route was soon superseded. Spices, instead of being shipped to the entrepôt markets of Egypt on their way to Venice, Genoa and Marseilles, always with a nice little profit for the powers of the day, were sent without stopping on a longer journey which was much safer. During this period the Turks, who had controlled Constantinople since 1453, turned the Mediterranean into a blind alley, or these conquerors did not follow the magnificent maritime

PREFACE

tradition of the Arabs who had done so much to develop trade. Then there was a change in the world's centre of gravity, which shifted westwards as Portugal, Spain and England each became the great power of the day and mistress of the seas. The Mediterranean was abandoned and left to the mercy of barbarian pirates who infested it for centuries.

The route to the Indies thus altered its course, and yet those who held the Mediterranean passage remained on the alert. In 1507 Venice had under consideration a plan for a canal; after that France with astonishing persistence studied the question of a direct route to Asia through either Syria or Egypt. Even when she lost India to England in the eighteenth century this preoccupation never quite disappeared. On the contrary it was she who initiated the expedition to Egypt which, thanks to the prestige of Bonaparte, gave France a favoured place there which she still enjoys.

When he visited the isthmus Bonaparte contemplated the traces of the ancient canal of the Pharaohs, and instructed his chief engineer, Lepère, to report on the possibility of following their example. Later Ferdinand de Lesseps, as a young consul at Alexandria, chanced upon Lepère's plan which had never materialised, and he became obsessed with the idea. It was his amazing adventure, his transposition of the Arabian Nights Entertainment into the reality of modern times, which ended in 1869 in the dazzling inaugural fête which the Near East recalls to this day. Owing to the genius of this man, whom we still refer to as 'The Great Frenchman', the Mediterranean regained its youth after four centuries of neglect.

But how the world had changed! The great power of the day was no longer Venice, Portugal or Spain. England was the undeniable ruler of the waves, and already fretting because an international highway which was not her work had dared to be established, and was even being administered without her. With her usual political skill, however, she eventually gained control, and in grand style. By buying up 177,000 shares from the Khedive Ismail in 1875 she penetrated into the company,

PREFACE

where she was graciously received by that clever politician, de Lesseps. Ten English administrators were appointed to the board of the canal, and there they are still. Their presence in a company which is legally Egyptian though actually French has always constituted an indissoluble link in the friendship between France and Great Britain. Actually the part played by the Suez Canal in the Entente Cordiale at the beginning of the twentieth century was, as we realise today, of prime importance. England was thus firmly established on the road to India, which has always been regarded as the axis of the British Empire. A few years later in 1882 she occupied Egypt, and inaugurated a period which was destined to last until 1914 during which she remained in uncontested command of the Mediterranean.

Meanwhile France, relying on the strength of Great Britain in their joint resistance to German ambition, altered her own centre of gravity to the western Mediterranean. In agreement with London she renounced all political aspirations in Egypt, and concentrated her attention on North Africa — Tunis, Algeria and Morocco — which was soon to become the keystone of her empire.

English policy, always logical and based on a thorough knowledge of the conditions of sea power, consisted in warding off the route to India — the route to the Indian Ocean if you prefer — any nation that she did not trust. France after the agreement of 1904 was no longer a rival. Italy sailed along in the wake of Great Britain, and in any case was associated with France in a policy of virtual resistance to German domination. Russia, after the Russo-Japanese war and owing to the good offices of Delcassé, was on friendly terms with the Western powers. A balance of influence was struck between London and St. Petersburg in that part of Asia which separates the Black Sea from the Indian Ocean. Germany herself had been checked in her *drang nach Osten*, and her ambitious project for a railway to Baghdad came to nothing since the terminus at Basra remained in the last analysis under English control. Never was a difficult

PREFACE

political situation more skilfully handled, for if the Mediterranean did not become a British lake it did not become the *mare nostrum* of anyone else either. Unremitting vigilance was needed, a marvellously supple policy, and amazingly ingenious methods and expedients. The problem was finally resolved, however, and the Mediterranean highway as well as the Cape route still remained under British influence.

The strength of Great Britain in world affairs is at once precarious and the result of sheer persistence. This empire relies on sea communications with its dominions and colonies scattered all over the globe, and therefore its equilibrium must be adjusted ceaselessly to meet constantly changing conditions. It is not a rigid structure, conceived and set up once and for all, but rather a piece of work that must always be renewed, a 'continuous creation', to quote one of Poincaré's brilliant expressions at the time of the Treaty of Versailles. It is also an association of principles, principles which do not change, for without them the system — if we may use the term for anything so flexible — simply would not function.

First it is essential to maintain a certain number of pillars to support this immense political community, call it 'Empire' or 'Commonwealth' as you will. These pillars were yesterday — and they may still be tomorrow — India, Australia, New Zealand, South Africa, Canada, and in a different way Singapore and Hong Kong. Free access must be maintained between the various parts of the system. This has been achieved by establishing bases and waterways to assure the liberty, or at least the security, of imports and exports, in short international trade. As a result the small, overpopulated island in the North Sea has managed to live and play the role of a unique connecting link between the nations of the world. Furthermore, to make the plan completely workable there must be a stable English currency, recognised on the international money markets and generally accepted as a means of exchange.

In the nineteenth century England considered India and the Indian Ocean as a sort of haven for her empire situated outside

PREFACE

the European area. At that time the routes uniting the East with the West seemed to be a factor of first importance. But from the moment that passage through the Suez Canal was free and reasonably secure, the Cape route lost much of its interest. Politically the latter was neglected. South Africa played the part of some *Ultima Thule*, attractive owing to her wealth, but isolated in the midst of immense, empty oceans that led nowhere. Whenever there was an international tempest it always gathered force in the vicinity of the Mediterranean.

The French believe that problems can be solved, and once they are out of the way one can sit back and plant cabbages, as we say, in other words go into retirement. How much wiser was old Luzzatti when he once remarked to me: 'My friend, there is no such thing as a solution!' The English, those inscrutable politicians, realise this. They are well aware that the best of solutions get out of date, and must continually be renewed and readapted to conditions which change with months and years. For them the state is a ship whose sails must be incessantly trimmed to make the most of every change in wind, of the moods of the weather, even of the slightest breeze which, to quote our La Fontaine, 'wrinkles the face of the waters'. The British statesmen know that they can never rest, that their work when it is done must be begun all over again, and always in new conditions and according to new principles.

Of this the recent transformation of the Mediterranean problem is proof enough, if proof were needed. After her victory over Napoleon, England having disposed of her most dangerous rival could have hoped to relax, but then de Lesseps and his canal created another grave cause of anxiety. So she protected herself, and as mistress of Egypt she seemed at last able to rest on her laurels. Then along came Germany, but England managed to hold her in check. Back she came again in 1914, supported by Turkey whose integrity London had so often guaranteed.

Then the problem changed in character. This same Mediter-

PREFACE

ranean, which England had made politically secure as a sea, became thanks to the submarine and the aeroplane the most dangerous passage of all. Her convoys were threatened, and although she kept military control over it, towards 1915 she certainly lost her commercial control. In 1915 and 1916 she was obliged to send her supplies another way. Looking down from wherever he is in the skies, Vasco da Gama must have smiled at the rejuvenation of the itinerary which four centuries earlier he had initiated with such skill and daring. The journey round the Cape was long and its ports were badly equipped. From 1917 onwards England preferred speed to security, so her convoys in spite of enormous risks returned to the old highway.

England gained the most, I think, from the victory of 1918. France played an equally important part in the struggle and her sacrifices were heavier, but England emerged from this test stronger than ever. There was no enemy to threaten her on the road to India. Germany was disqualified and out of the running, and Russia had retired into Asia where she seemed very far away. It was easy enough for the British Government to have the Treaty of Versailles confirm their privileged position as guarantor of the freedom of the Suez Canal. If any further difficulties existed it was merely with an Egypt whose nationalism was exasperated by the presence of too many foreign troops on her soil.

For more than fifteen years England negotiated, discussed and tried every method to reach agreement, admitting that times had changed, that the age of Arabi Pasha and Lord Cromer had passed, that Egypt had developed owing to her contact with the West and was now ready for a different international regime. Egypt was no longer a vassal state, owing allegiance to Turkey as she had been at the time of Mahommed Said or the Khedive Ismail. A 'protectorate' as in 1914 was no longer appropriate, for now her sovereignty evidently had to be recognised and respected. But Egypt was very much mistaken when she thought that England — that the West — could ever

PREFACE

lose interest in the great route to Asia. True, the canal might be on Egyptian territory and the Egyptian boundary line might be 200 miles to the east, but there could be no question of the canal zone ever becoming exclusively and effectively under the sovereignty of the Government of Cairo.

England never for a moment intended to release her hold on the isthmus. One need only recall the reservations which she appended to the Kellogg Pact — a ridiculous treaty, in my opinion — according to which the powers solemnly undertook not to resort to war. The English signature, the only honest one in the list, stipulated that in the case of certain regions which were essential to the security of her communications, she was not willing to commit herself. In these circumstances, although the 1936 Treaty with Egypt modified the terms of occupation it did not bring any basic changes. Under the form of an alliance between equal partners, England remained, if not necessarily in Egypt, at any rate in the canal zone. It was on the strength of this alliance — and what an alliance it proved to be — that she was present in this part of the Mediterranean during the whole of the second world war. It was as Egypt's ally that she defended the approaches to the Nile and maintained the security of the isthmus. It is still as an ally, and in spite of Egyptian protests against a treaty that remains in force, that she is there today.

During this second world war, which was like the first and yet very different, the Mediterranean route was no longer what it had been before. The progress of the powers of evil was such that the means of destruction had increased tenfold, a hundredfold, until it had become impossible to use the Mediterranean route commercially. Gibraltar held fast; Malta, bombarded scores of times but always able to resist, also held fast; and Egyptian territory was never violated. Today we know that out of ten convoys destined for Malta, it was seldom that two ever made safe journey to port.

The British fleet retained control of this narrow and dangerous sea. It could no longer be used as a supply line, however,

PREFACE

so as in 1915 the convoys bound for England went round the Cape. England did not give up the Mediterranean, but she now saw it in a different light. As it was impossible to keep it open and workable, she contented herself with guarding its two bottlenecks, Gibraltar and Port Said. Egypt became the advanced glacis in the defence of India, and the distributing centre of the military forces destined for the struggle in the eastern Mediterranean and western Asia.

In this conception the Indian Ocean retained its full importance and England succeeded in dominating it. Even the fall of Singapore did not seriously affect this control, nor yet the advance of the Japanese towards Bengal. Relations with western Europe and India were thus guaranteed, and communication could even continue with Russia by way of the difficult roads through Persia. Egypt also kept in touch with England and the United States, either dangerously through the Mediterranean, or slowly but surely around the Cape, or by the Gulf of Guinea from which a liaison of motor lorries ground across the trails through Chad and Khartoum. The system was complicated and difficult but it worked, and victory was around the corner.

Dear old England, how magnificent she is! Magnificent because she is always there, courageous, ready to put her iron back into the fire a hundred times if need be, to take up new positions a thousand times, endlessly ready to change her drills, her methods and her expedients. She is always working with the same aim in view, with the same persistence towards the same goals which are necessary, not to her ambition, but to her very life. Yet she can never relax, for although her work is always successfully achieved it always has to be done over again. The problems are always the same, but they present themselves from a new angle and demand an ingenuity, a vitality and a political genius which must always be on the alert.

We are now in 1947, and the Mediterranean is still a British problem. Let us run through the prophetic dates which punctuate this amazing story:

PREFACE

1498   Discovery of the Cape route.
1588   Defeat of the Great Armada, and beginning of England's supremacy in world affairs.
1798   Reappearance of France in the East as a menace to British supremacy in India.
1869   Opening of the Suez Canal in spite of Palmerston by a Frenchman, Ferdinand de Lesseps.
1875   Purchase of shares in the Suez Canal.
1882   Military occupation of Egypt.
1915   Germany and Turkey reach the boundary of the canal zone and threaten to invade it.
1918   Victory, which one dared hope would be lasting.
1936   Anglo-Egyptian Treaty.
1940   Germany regains the offensive; Egypt threatened; difficult defence of Syria and the delta.

The same old difficulties reappear under a new guise, and once again England undertakes her labour of Sisyphus.

Let us see how this eternal problem of the Mediterranean presents itself today and for the future. Would the route through the Suez Canal still be an ulterior motive for Britain if India were to leave the Empire? There are three answers in the affirmative.

Firstly, Australia realises that she always has the Cape route to fall back on, but the direct route through the Mediterranean is her closest link with the Mother Country. New Zealand shares this opinion, but being geographically nearer to the Panama Canal is less insistent. However, England feels encouraged by the anxious preoccupation and support of her two dominions far away in the Pacific.

Secondly, the existence of an unlimited source of petrol in Iraq, Persia and Arabia has led Britain to set up in the Near East and western Asia an economic fortress which has become almost vital to her security. At the beginning of the present century the British fleet abandoned coal in favour of oil and petrol. Consequently British statesmen must now not only make

PREFACE

sure that the way to India is kept open, but also that Iraq, Syria, Persia and Arabia do not fall under any foreign influence that would endanger this new link in the imperial chain. Suez doubtless is not indispensable in this respect since the outlets of the Imperial Oil Company's pipe-lines are situated at Haifa and Tripoli. Nevertheless this complex group of countries is so closely bound up with the fate of Egypt in their common Mahommedan allegiance that they cannot be considered separately.

Finally, there is still India to be considered and also the Indian Ocean. The British have now undertaken to let India govern herself and the last of their officials will have left by June 1948. The British play the game well, and they always keep their promises when they recognise that a country is determined to obtain its independence. Perhaps this time they may keep their promises even better than the interested parties could wish. This is an extraordinary state of affairs and England has undertaken an extraordinary risk. Who knows? Probably she argues that India, such as she is and such as England has made her, cannot, will not even wish to do without the British. It is quite possible that in one form or another Britain will be recalled in new conditions in which her presence will no longer be imposed but requested. This would leave India still a member of the British community. Quite apart from her domestic government, India must now consider her international position. This immense country of 400 million people, racially so old but nationally so young, will wonder whether it really is ready for complete international independence. No doubt some form of protection will be necessary and then British power will still remain in the Indian Ocean.

Up to the present the Indian Ocean has been a British lake, with no direct threat of foreign intervention. From the British point of view it has been so completely safe politically that a very small naval force has been sufficient to guard it. Japan has always kept far away to the north of Singapore or Hong Kong, and Russia to the north of the Himalayas, Afghanistan,

PREFACE

Persia and the Bosphorus. This integrity must be maintained today, not against Japan which is out of the running, but against the U.S.S.R., which is enormous, dynamic, ambitious and shut in between narrow seas to which she naturally desires an outlet. Russia wishes above all to reach the Indian Ocean, but she finds that almost all the countries bordering on it are under British influence. So the question of the sea lanes leading to it is again to the fore, and more urgently than ever.

The Cape route is more useful today than it was, for during the two wars the equipment of its ports was improved, and in any case oil-burning ships can now make longer journeys without refuelling. South Africa on the other hand has developed, and, important to note, many Indians have settled on the East African coast. The time seems to have come to affirm the presence of the British. This consideration may have influenced the decision to send the King on a visit to the Dominion, a visit which proved to be not only symbolic but spectacular in its welcome.

Nevertheless, the Mediterranean route, which after all is the quickest and passes alongside regions which definitely are under British aegis, still cannot be abandoned. Traditionally it is a waterway, but now it is also an airway to which Egypt is the key. Today it is not only a case of the security of the canal, but also of freedom of transit by both sea and air in a much larger area in which the isthmus and the delta are essential factors.

We thus see the problem in its wider aspects. It is no longer simply 'canalist' as the Khedive Ismail once described it, it is Egyptian. It is not only Egyptian but Syrian, for Syria lies on the shortest air route. So it all blossoms out into a Mediterranean problem. Since modern warfare is carried out on a much larger scale, the canal can no longer be defended from its own small territory. To defend the isthmus and the delta efficiently in future it may be enough perhaps to control Palestine and Cyrenaica in strength, and towards the south Khartoum or Kenya. With this triangle of vantage points in her possession

PREFACE

England could feel reasonably secure, without strictly speaking occupying Egypt itself.

The Sudan and Kenya now assume an importance which it took the genius of Cecil Rhodes to foresee, but which up to the present has never been fully appreciated except in well-informed circles. One can hardly over-estimate the value of the route overland from the Gulf of Guinea, which was used during the last war to bring supplies up to Egypt. On the other hand and more than ever before Britain must see that order is maintained in North Africa. It was there in 1942 that the first offensive was started which eventually led to the defeat of Germany on the continent. In a struggle between the continent and the totalitarian states it again would be there, based on the Mediterranean peninsulas, that the Anglo-Saxon powers would organise an offensive for the reconquest of Europe.

The Mediterranean figures in all these combinations, and even as a central factor. Far from considering it as a backwater away from the nerve centres from which the destiny of the world is directed, it must be regarded today, yesterday and always, as a unique and special area meriting special study. It is this study that we have attempted in this book. We have not tried to make it topical, for events change so rapidly that constant revision would have been necessary. Instead we have dwelt on the lasting aspects which make the Mediterranean a world centre where something is always happening, and to which attention invariably returns.

The Mediterranean was great in ancient times when the civilisation of the Greeks and Romans, although marvellous even then, was only a rudimentary technique. Yet it was left behind when the modern industrial revolution took place, for those in the vanguard of progress were no longer the people of the Mediterranean.

We have tried to show how these people have managed to adapt themselves to new conditions of production that are hardly suitable for a territory which is almost devoid of coal. We have also tried to show that their traditional qualities,

PREFACE

founded since the time of Homer on initiative, intelligence and ingenuity, will always have a place, and that the Mediterranean will always form part of the Western world. The Mediterranean is a civilisation, and it is also a highway. Finally, it is one of the regions in which political tempests arise, gain momentum and burst forth. Today it is more necessary than ever to know the Mediterranean, and above all to understand it.

Paris
November 1947

INTRODUCTION

A General View of the Mediterranean

WHEN we consider the face of the world, we easily discern several great civilisations, all geographically distinct, and each with its own way of living and its own conception of production and social relations. The Mediterranean region expresses a conception which is typically European, whether it be of the individual and the family, or of production and trade, or of life itself. The Mediterranean is at the opposite pole from the industrial system of North America, but on the other hand it has played such an important part in the formation of Europe that no study of the Old World would be complete without it. Its influence goes still further, for without the Mediterranean we cannot understand the relationships between Europe, Asia and Africa, the contrast between the old and the new continents, even the co-operation between East and West in the development of the civilisation of the human race. The subject is great enough to be intimidating no matter from what angle we approach it, and yet we must take courage and face it undismayed.

I

The Mediterranean consists of a deep gash in the crust of the earth, no less than 2250 miles long reckoned from Gibraltar to the Syrian coast. If one were to trace the same distance elsewhere on the map, one could go west as far as Greenland or east as far as Delhi (Fig. 1, p. 26). Yet it is just a narrow sea, only 470 miles wide between France and Algeria, and 500 between the Libyan coast and Salonika. The other seas in this region also are restricted in area, the Adriatic being only 500 miles in length and 125 at its maximum width, while the Black Sea is less than 700 by 400 miles. Contrasted with the enormous

INTRODUCTION

size of the oceans, the indentations of its shore line are geographically one of its most interesting aspects, and at the same time of the greatest consequence.

As a sea its boundaries are quite definite. In the Straits of Gibraltar which separate it from the Atlantic, even the most weary traveller is impressed with the fact that he is passing from one world into another. At the Isthmus of Suez the know-

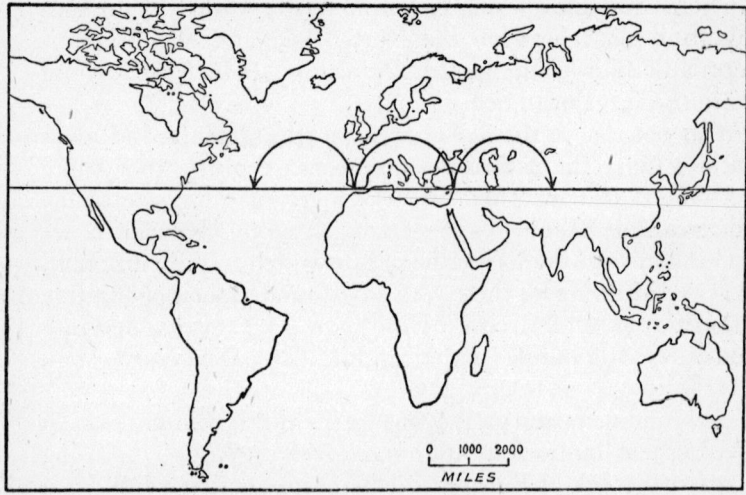

FIG. 1. POSITION OF THE MEDITERRANEAN ON THE MAP OF THE WORLD

ledge that one is crossing one of the great frontiers of civilisation is even more momentous. E. M. Forster, in *A Passage to India*, has admirably expressed this emotion which is felt by most travellers: 'It is in the Mediterranean that humanity finds its norm, and when mankind leaves this exquisite lake, whether it be past the Pillars of Hercules or through the Bosphorus, it approaches the world of the unknown and the monstrous. It is, however, by the southern route through Suez that one gets to the most foreign region of all.'

The boundaries of the Mediterranean zone are almost as clearly defined, for both climate and living conditions soon change once we penetrate inland from the coast. Almost

INTRODUCTION

immediately we are either faced with mountains, or are already in the desert.

The entire Mediterranean is really not very large, 2250 miles long and 1,160,000 square miles in area. We soon realise its limitations when we travel from one end to the other, either by boat or more especially by aeroplane. After all it is only the transition between temperate Europe and tropical Africa. We have here something which is unique in the world — a delightful halting-place between the North Pole and the Equator. There is nothing of the kind in Asia, nor yet in America, not even in the Caribbean Sea.

Let us not ignore the fact that since the fifteenth and more especially since the end of the nineteenth century, everything has changed in scale. We can no longer take the size of the Mediterranean as our basis of comparison. Henceforth we must reckon with the size of the oceans which is quite different, and very disturbing for those who have founded their civilisation on the harmony of Greece. Unless we take this change into account, we cannot fully understand the critical period through which Europe is passing today.

This prolonged deep gash opens up a natural communication between the Atlantic and the Indian Oceans. Also, by splitting the massive block of Europe and Asia, it acts like a link, as a hyphen, between East and West. We have here a geographical phenomenon far exceeding any local framework, and without exaggeration it is of world-wide consequence and proportions. We are not considering just the finest of lakes but a highway, an imperial thoroughfare, and this places our study on an intercontinental basis.

Let us begin by ascertaining the exact position of the Mediterranean as to latitude and longitude (Fig. 2, p. 28). In latitude, Tripoli on the most northerly part of the African coast is about the same as Louisiana or Shanghai. Trieste on the most northern point of the European coast compares with Montreal and Vladivostok. Thus the Mediterranean lies along the same latitude as regions such as the southern Mississippi which is

INTRODUCTION

much hotter, or Canada which is much colder. This emphasises the contrast between the temperate Mediterranean and the climatic extremes of North America.

In longitude Gibraltar corresponds to the west coast of Scotland; Algiers with a line running through Paris, Barcelona

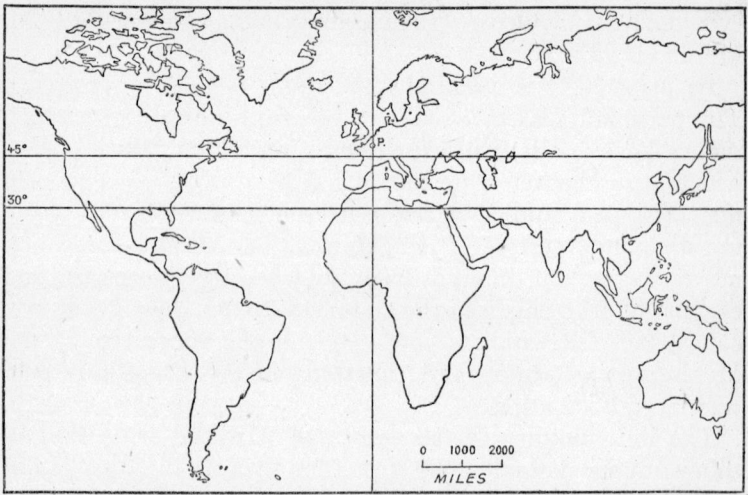

FIG. 2. THE LATITUDE AND LONGITUDE OF THE MEDITERRANEAN

and Kotonou; Alexandretta with the White Sea and Lake Victoria; and Batum with the Mozambique Canal. These comparisons are purely European and African, but they indicate the importance of the Mediterranean's position as a link between the two continents. From the point of view of uniting France with Africa, nothing could be more significant than the line running through Paris, Marseilles, Algiers and Kotonou. As a result this sea, quite regardless of its size, cannot be conceived as a basin closed to the outer world.

11

Those who visit the Mediterranean are invariably impressed with its unity. Everywhere it is the same, for the shades of

INTRODUCTION

difference here are less important than the resemblances. Yet this unity is the result of aggressive contrasts—sea and mountain, sea and desert, sea and ocean! In these respects the Mediterranean is very different from either central Europe, or the high tablelands of Asia, the deserts of Syria or Sahara, or even the Atlantic Ocean. In describing it as anti-desert M. Paul Morand is quite accurate.

Its personality is essentially the result of these oppositions. This personality is expressed in its geographical structure which relates back to its geological formation; also in its climate, which is so characteristic that it has been taken as a type in meteorological studies; again in its atmosphere which is made up of colours, perfumes, temperatures and a radiance which makes this region different from anywhere else on earth, and essentially different from the countries surrounding it. As one can always tell within a couple of miles if one is still in the Mediterranean atmosphere, no study of boundary lines ever was more fascinating.

The individuality of this area is determined by a host of things: by specialised conditions of production; by a particular type of trade; by a way of living, which forms a certain type of man who remains quite distinct from those who emigrate; by the political nature of the countries on the shores of this sea; by its role as an international highway, although it is so narrowly shut in at each end; and finally by the very fact that a whole civilisation has been born and nurtured here.

It is quite easy to pick out this civilisation on the map, or at least to see how it radiates from its centre. In the same way it is easily traced in the history of the human race, in which it has formed an essential phase linked up with a certain form of production, with the tool and with individual thought.

M. Paul Valéry's observations on this theme are decisive: 'The edification of human personality and the development of an ideal of the most complete and most perfect man have emerged and been brought to fruition on these shores. Man, the measure of things; man, the political element, the man of business; man, the legal entity; man, the equal of the man created in the likeness of God and considered *sub specie aeternitatis*. Here we have creations which are almost entirely Mediterranean, and there is no need to dwell on the immense effect they have had on the rest of the world. Whether it is a question of natural law or civil law, every kind of law has been defined by learned Mediterranean minds. It is here that science has freed itself from empiricism and from the shackles of practice, that art has risen above its symbolic origins, that literature has gone forward clearly into new and different fields, and finally that philosophy has embarked on almost every possible way of contemplating itself and the universe.'[1]

This definition presents a problem, that of the place that the Mediterranean civilisation will be likely to hold in the future. The world of tomorrow, will not necessarily, not even probably, be based on the same principles or radiate from our present centres. Already we see the machine supplanting the tool, mass production opposed to craftsmanship, the group or the gang taking the place of the individual worker, and collective organisation replacing personal initiative. Ford seems to be confronting Ulysses, the patron of ingenuity and of *debrouillage*. Even Europe, which is so largely situated in the Mediterranean, is involved in this menace.

III

The contrasts to which we have alluded are one source of the contacts which exist, firstly, between the regions adjoining the countries bordering on the Mediterranean; secondly, between the different continents; and thirdly, between the different civilisations. In this respect the role of the Mediterranean is of prime importance. The very fact that a sea should be in this particular place, that it should have this particular form and these particular characteristics is essential for the Old World. Actually it is quite as essential for Asia and Africa as for Europe,

[1] PAUL VALÉRY, *Variété III*, p. 264.

INTRODUCTION

for a breath of fresh air passes this way and refreshes everything as it passes.

France would not be herself without her Mediterranean aspect. As one realises she faces three ways: towards the Atlantic, towards the continent and towards the Mediterranean. Owing to her Atlantic façade she is Western in outlook and atmosphere, looking out on to the world at large. She maintains her relations with non-European communities such as the British Empire, the United States and Latin America — in fact with the whole of the Western section of our civilisation. On her land side, however, she belongs physically to the continent, is part of its very being — a bond which England does not share. The whole of the eastern section of France — Burgundy, Alsace-Lorraine and the Free County — is bathed in the atmosphere of central Europe which is not strictly Western. Hence, France is continental, so typically European indeed that it would be difficult to conceive of her except as an indissoluble part of the old continent.

Finally, her Mediterranean frontier brings France into direct and immediate contact with Africa, Asia and the Far East, in other words with the most illustrious past of the human race. She thus goes back to the civilisations of remote antiquity and to practices as old as the Stone Age. Where one contemplates the terraces erected by the cultivators of the south, one has the impression that these people belong to an infinitely venerable tradition, for their very patience is touching. They link us with Asia, a link which the Anglo-Saxons appear to have severed.

To these various contacts one may attribute France's special qualities, in that she can look into the future and back on the past, and at the same time be attracted by progress though remaining attached to tradition. Herein lies a double attraction, but it can also be contradictory — hence the role of the Mediterranean. In this country it is divided into the Mediterranean Midi and the south-western or Atlantic Midi. It would not be too much to say that the former is the only genuine or true Midi. What France would be without it, what

INTRODUCTION

she is with it, and how much she owes to this coast with its northern and southern attractions, is a most absorbing study. I shall hope to throw some light upon it without becoming too deeply entangled myself.

For her part Europe owes one of her contacts with the East to the Mediterranean, but only one, for by land her communications with the Far East and Mongolia are through Russia, across the great plains of northern Europe. This difficult route is subject to many bewildering transitions, so that although Europe penetrates deeply into Asia across Russia, Asia also penetrates much further than one imagines towards the West. On the other hand the contact which is made via the Mediterranean is clear cut, with definite frontiers. The Mediterranean may be deeply committed in East Africa and Asia, yet it remains part of the West, forging a much more effective link than Russia does, for one never really knows whether Russia is Asiatic or European.

The reason is that one of the principal maritime routes of the globe passes through the Mediterranean. Europe's sea communications with the other continents follow three main routes: one network crosses the Atlantic towards the New World; another goes round the Cape towards the Indian Ocean and Australia; and the third goes through the Suez Canal towards India and the Far East. The latter route, which was revived by Ferdinand de Lesseps, must be maintained at all costs, more especially as its security is threatened once more. In point of fact the Suez Canal today presents a political problem that is so serious that it can hardly be exaggerated.

Nor can we speak of the Mediterranean without conjuring up the destiny of the whole of the West. The white race has always sat astride Asia and Europe, but it was the European section that created the Western civilisation. In their opposition to the Persians, the Greeks of antiquity already were Westerners in the true sense of the word. We Westerners should all be making pilgrimages to Marathon!

What I have said is incontestable, but we must not overlook

INTRODUCTION

the fact that a new section of the white race has been formed in recent years on the other side of the Atlantic. This is a historical fact of outstanding significance, for a new province of our civilisation is developing outside the boundaries of Europe in the British Empire, in the United States and in Latin America. As a result the Mediterranean zone is now set aside for the second time. In the sixteenth century the inter-continental route joining Europe and Asia ceased to pass through the Mediterranean, and although in the nineteenth century it again became the main thoroughfare between East and West, the twentieth century is challenging this monopoly, either by reviving Vasco da Gama's route around the Cape, or by moving the centre of gravity of the planet further towards the West. The Mediterranean is in danger of becoming out of date in an industrial age when the great centres of production are growing up elsewhere.

The Mediterranean is capable of adapting itself to new ideas, as has been fully demonstrated by the admirable initiative with which it has modernised many of its methods. Perhaps it is better for it to remain true to itself, for the world may need the values which it stands for. In these conditions it may yet play a leading role as the international economic link which de Lesseps inaugurated, and also as the guardian of a tradition of individualism and ingenuity, mingled with patience and frugality. America, so far removed from the heritage of Greece, is now abandoning these ideas, at the risk of directing civilisation along paths which may be magnificent but which are still unknown.

IV

In this survey it is the economic civilisation that I especially intend to study. A geographical preface seems to be indispensable, for there are few regions where the structure of the soil, the climate, even the sea itself, have a more direct bearing on human development. We shall have no difficulty in appraising

INTRODUCTION

the sense of values of the Mediterranean people, once we have ascertained the various races from which they are descended. There is something emotional in this collaboration between nature and humanity, and nowhere is it more emphatic than it is here. It is essentially in the Mediterranean that one can say that nature has made man. Man in his turn, however, has worked with nature and by bringing life to the waste places has enabled civilisation to carry on. This recalls Bacon's wise saying, *Ars, homo additus Naturae*.

I am well aware that my subject has already been handled in masterly fashion by some of our leading geographers, by men of the stamp of Vidal de la Blache first of all, but also by Jean Brunhès, Jules Sion, Maximilien Sorre and several others. I have learned a great deal from their work, and do not pretend to have thrown much new light on the subject. I like to think that the Mediterranean has inspired many writers, and among them some of first rank. One must see it through their eyes. I have quoted in particular from certain contemporaries whose observations seem to be both penetrating and profound. André Gide, the Comtesse de Noailles and Paul Morand will forgive me for turning to them so often. The work of the late Paul Valéry has proved invaluable.

Literature, poetry, painting, and even an analysis of odours and perfumes are indispensable to geographical comprehension. That keen thinker, Doudan, once said: 'All the secret harmonies of creation should at least be indicated if not described' — an excellent rule for anyone observing nature. Now such a task, I think, is best undertaken by writers, poets and painters. The world of imponderables belongs to them. Therefore I have turned to them for guidance, as well as to our leading geographers.

May I add that I myself have many times travelled in the Mediterranean region, either crossing the sea itself by boat, flying above it by aeroplane, or wandering along its shores. These journeys have left a most vivid impression, and this is my excuse for being so bold as to attempt so great a subject.

THE MEDITERRANEAN

CHAPTER I

GEOLOGICAL ORIGINS

THE historical age of the Mediterranean contrasts sharply with its geological age. Historically this is one of the oldest regions in the world, and geologically it is one of the youngest.

Situated almost entirely in the zone of the Alpine foldings, but directly affected by much earlier consolidations in the vicinity, the Mediterranean depression is not simply a local phenomenon. As its existence is linked up with some of the great events of the evolution of the earth, it must be regarded as part of the surrounding structure.

It is the Tertiary era which has chiefly left its mark on the geological character of the Mediterranean, but to appreciate this one must also consider the double framework of two primary stable masses. These masses are very ancient and worn, with little or no trace of folding. On the one hand we have the European framework — the Euro-Asiatic mass which constitutes the underlying core of Europe; and on the other hand the African framework — the Indo-African mass which serves as the foundation of the desert. This is the solid and stable structure of the Old World (Fig. 3, p. 38). These orogenic efforts, which were violent and repeated although slow, were followed in fairly recent times, in fact almost at the appearance of man, by the formation of two opposing masses which served as buttresses. Out of this has arisen the structural contrast of the two tablelands of the Mediterranean region. Between them lies an immense gap with bold well-defined peninsulas and abysmal gorges. This daring architecture is the result of the juxtaposition of early and recent forms, of dislocations and subsequent sinking. Such is the character of the region.

'The geologists, those historians of silent revolutions,' writes Paul Morand, 'show us that the Mediterranean has arisen in

THE MEDITERRANEAN

reaction to the deserts like a liquid response to endless drought, or a majestic signpost pointing in the opposite direction... The Mediterranean is the anti-desert.'[1]

Let us now place the Mediterranean in its proper relationship to Asia. The two archaic masses forming its framework are the

▩ Euro-asiatic group   ⌒ Alpine folds
▨ Indo-african group   ∥ African rift valleys

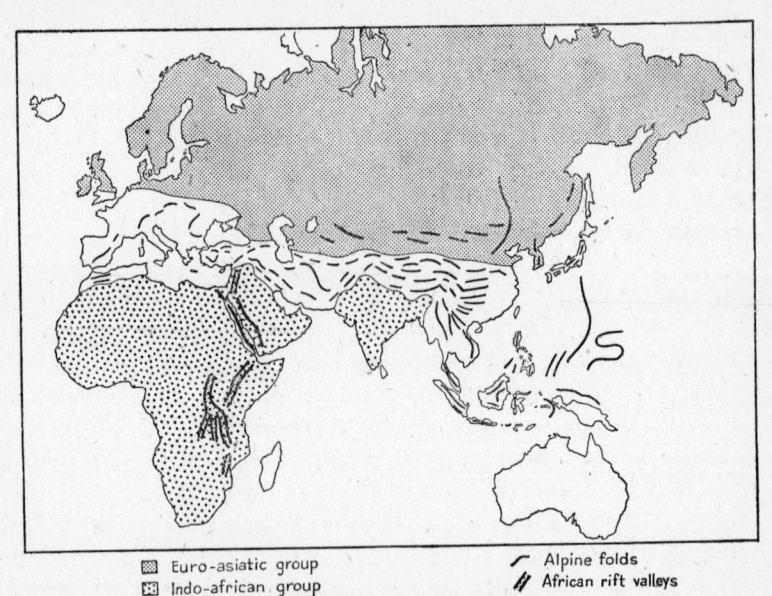

FIG. 3. THE STRUCTURE OF THE OLD WORLD

continuation of two Asiatic masses of the same nature: the Sino-Siberian massif to the north, and to the south the Indo-African massif which is the Gondwanaland of the geologist, Suess. This accounts for the underlying geological unity between Asia, Africa and Europe. The central Mediterranean — the Tethys of the geologists — is said to have already existed in the Secondary era, and possibly even at the end of the Primary era. It was then much more extensive than it is today since it separated India from the Sino-Siberian massif, just as the Mediterranean now separates Africa from Europe.

[1] PAUL MORAND, *Méditerranée, Mer des Surprises*, p. 13.

38

GEOLOGICAL ORIGINS

This depression is counterbalanced by the Alpine folding of the Himalyas rising up in the centre of the continent of Asia. This mountain chain is the source of great rivers, and indirectly accounts for the immense alluvial plains of India and China where the civilisation of the human race was nursed into being. An obstacle thus arises in Asia, but the maritime communication remains in the West. In Asia communications do not run north and south, but between east and west. The Mongol invasions unfurled east and west, and it is still in this sense that a link has been established between India and the Mediterranean and between India and the Far East. The very fact that it provides this link emphasises the importance of the eastern Mediterranean, indeed of the Mediterranean itself.

II

Apart from certain exceptions, the Primary formations lie outside the Mediterranean region. They are attached to the Euro-Asiatic or Indo-African platforms which geologically are not part of the Mediterranean region (Fig. 4, p. 40).

A few rare traces of Primary formation remain here and there in the Mediterranean zone, the remnants of the ancient hypothetical Tyrrhenian continent. These crystalline rocks date back to a period before the coal age, being more or less transformations of the results of ancient eruptions. Examples of this are the massifs of Catalonia, of Albères, the Maurès and Esterel, Corsica, Sardinia, Calabria, and the two massifs in the mountainous region of Kabyle . . . scattered witnesses only, but yet all grouped together in the western Mediterranean.

As for the highlands forming the framework of the two continents, we find high masses, solid and stable, which have been thrown up from time to time among the more recent formations. According to Suess they are the remains of the continent of the Altaides which stretched from Asia to the Atlantic. These remains consist of massifs which have been hurled forth, and of worn formations which are often capped

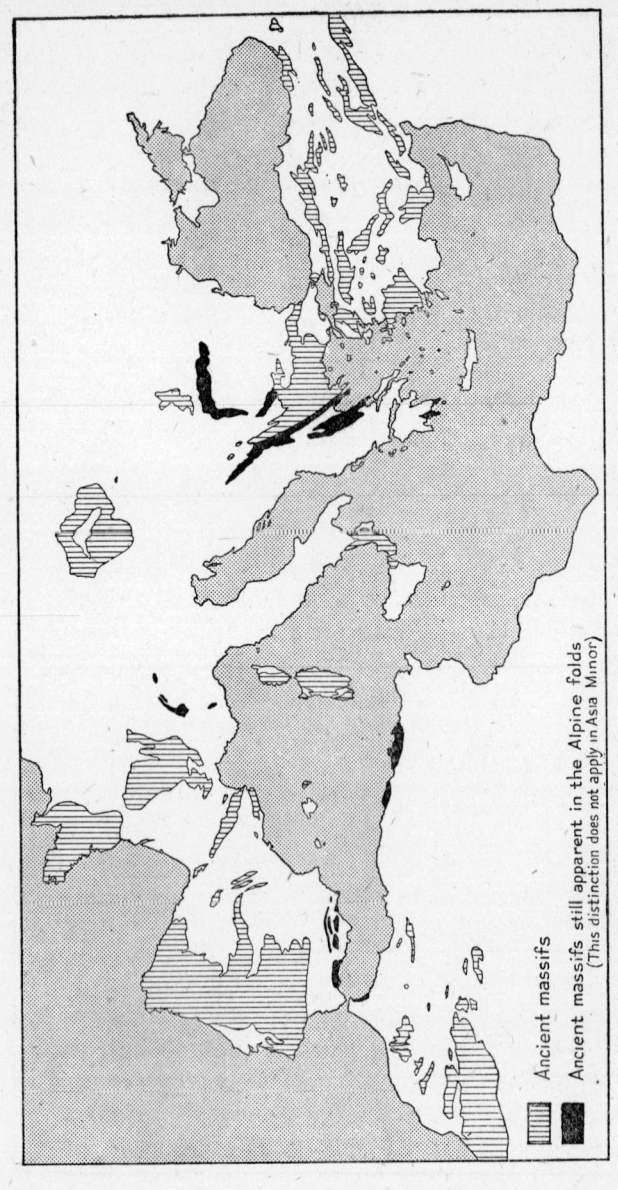

FIG. 4. ANCIENT MASSIFS IN THE MEDITERRANEAN AREA

Ancient massifs

Ancient massifs still apparent in the Alpine folds
(This distinction does not apply in Asia Minor)

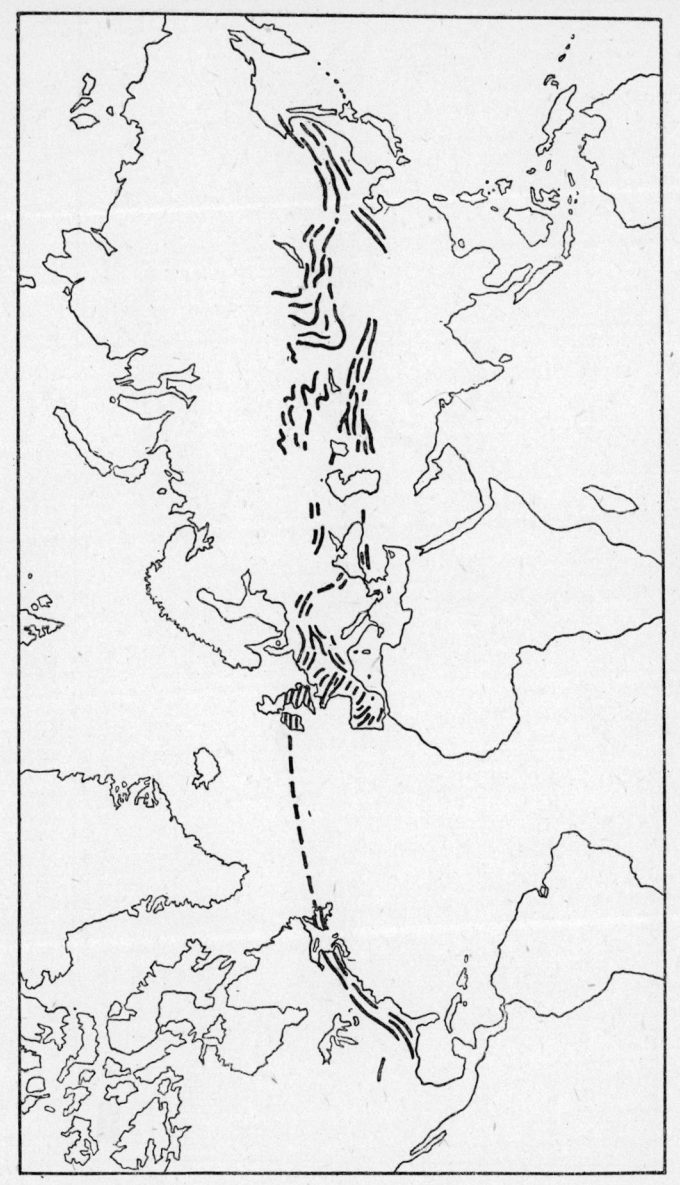

FIG. 5. THE HERCYNIAN CHAIN

with peneplains or tracts of land which are almost flat. Concrete examples are the Pyrenees which are the hub of the wheel; the massif of Rhodope; and high tablelands which are composed of sedimentary moraines such as the Mesetas of Spain, Oran and Morocco; and finally the high tablelands of Asia Minor. Note that these plains, although they are only a framework, still belong to the Mediterranean, and while preserving their own identity still are part of its atmosphere.

Although it lies outside the Mediterranean region we must mention the remains of the Hercynian chain, a formation dating back to the end of the Primary era, as without it it is impossible to understand the place that this sea occupies in the present economic structure of Europe (Fig. 5, p. 41). The learned term, Hercynian, which is used only by geologists, also has a definite meaning for economists. Hercynian conjures up the 'black countries' of the nineteenth century, where the steam engine was invented, and coal was produced as well as iron and steel. The term implies not only mineral wealth and metallurgical development, but also manufacturing power and consequent political influence. This must be borne in mind, if only to emphasise the fact that the Mediterranean is not Hercynian.

The Hercynian chain is an enormous wrinkle straddling continents and oceans alike. We find its most westerly point in the United States in the Sierra Comanche west of the Mississippi; the Allegheny Mountains belong to it and also their prolongation into Nova Scotia and Newfoundland. The Atlantic Ocean does not interrupt this phenomenon for the folding reappears in the British Isles, the Armorican massif, western Spain, the Ardennes, the Vosges, the Black Forest, the borderlands of Bohemia, and finally in the Harz Mountains (from which Suess derived the term Hercynian). The chain then continues eastwards through the Carpathian and Donetz Mountains, Russian Turkestan, the Altai Mountains as far as Manchuria, and finally through eastern Siberia to the Pacific. The chain has thus almost encircled the globe! We have followed an immense trail along which nearly all the industry of

GEOLOGICAL ORIGINS

the West was developed during the nineteenth century. But the Mediterranean is not part of it (Fig. 6).

From this one can easily perceive the role of the Primary formations in the constitution of the Mediterranean world. They make up the framework, but in this case the frame reacts

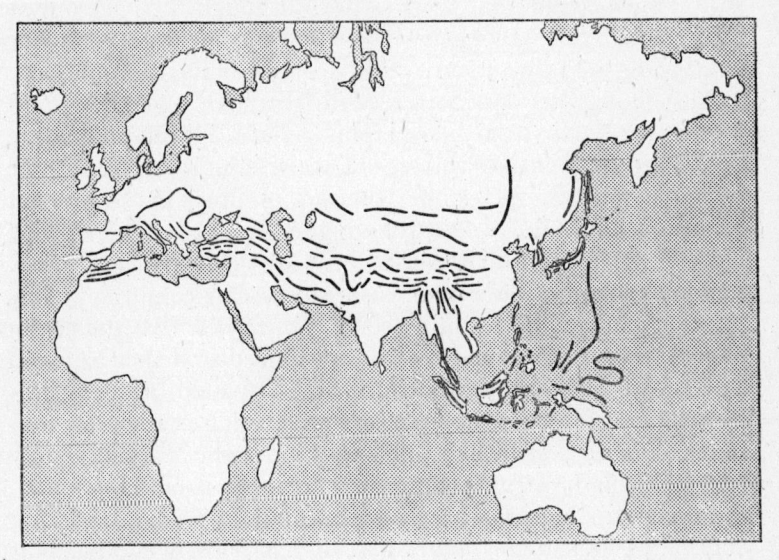

FIG. 6. THE ALPINE FOLDS AND THEIR ASIATIC EQUIVALENTS

upon the picture and without it the picture loses its true perspective. As for the Hercynian chain, it is its absence that is important, for without it the Mediterranean is not metallurgical and therefore has no heavy industry. These economic characteristics were apparent as early as the Primary era or in primitive geology.

It was at the end of the Primary or Secondary era that the Mediterranean appeared like a long narrow ditch stretching east and west, but much larger than it is today. Its coastlines have altered greatly as there was a tendency for it to fill in from the end of the cretaceous age, but even then it was a prefiguration of the present sea. One might say that the Tethys of earlier times which overflowed Europe towards America and towards Asia was an indication on a much larger scale of the basin which today is of interest to the entire planet. But it was between the Secondary and the end of the Tertiary eras that the great Alpine chains were formed. This line of intense crumpling extended across Asia to the Himalayas, and from there to the Sunda Islands and Japan in the Pacific. The Mediterranean undoubtedly owes its essential individuality to this world-wide repercussion.

A first series of folds resulted from the general pressure of the masses of the two continents. Violent explosions occurred, breaking up the surface in all directions into abrupt formations, as exemplified in the central Iberian chain, in the Pyrenees and lower Provence, with thrusts also towards the Caucasus. The pressure of the two jaws of the vice was then accentuated. This was the true Alpine phase when the Sierra Nevadas were formed as well as the Alps and Carpathians, the Balkans, with the concavity of the Alpo-Carpathian folding outlining the convexity of the African continent. Finally, lesser folds were annexed to the African framework, and these can be traced today in the mountains of the Rif, the Atlas and Apennine chains, and the arcs of Dinaro-Taurus and Anatolia (Fig. 7, p. 45).

These phenomena with intervening paroxysms succeeded one another over a long period. Thus the Pyrenees are older than the Alps, and the Alps are older than either the Apennines or the Dinaric Alps. The forms are as much younger, i.e. more accentuated, as the formation is more recent. In the Mediterranean zone it is the Alpine folding which has left its mark on the essential design of the structure, either through the great arc of the Alps which continues through the Dinaric Alps towards Crete and Asia Minor; or through the Betic Cordilleras which link the Alps at Nice to the massifs of Corsica and Sardinia to the south, or again through the Apennines and more to the north, with the Carpathians.

Wherever the Primary formation is stable and worn down to a hard surface, the Tertiary Alpine folding is on the contrary young, bold and contracted. There is close contact, actual mutual penetration, between the sea and the mountains, and as the latter are often under water they rise up from the sea like a precipitous wall. One of the most significant characteristics of the region is that from the sea one can discern the mountains

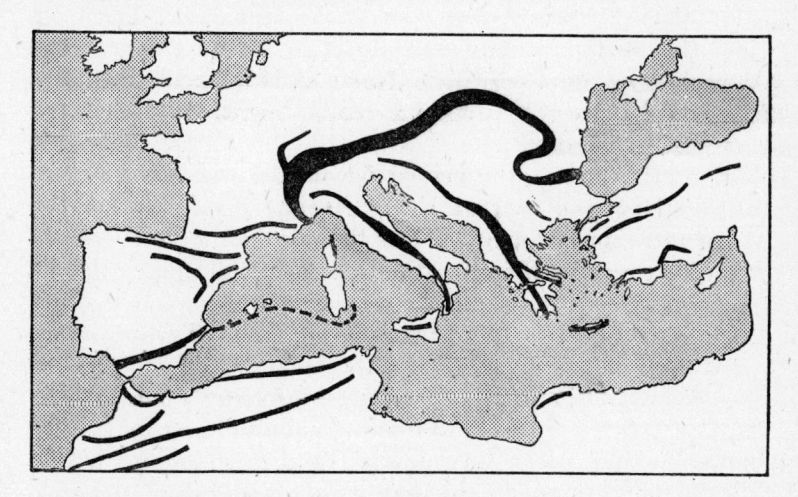

FIG. 7. THE POSITION AND DIRECTION OF THE ALPINE FOLDS

almost everywhere overhanging the shore or outlining the horizon. As a result the river beds are never very wide and there are almost no plains. As we shall see when we study the Mediterranean ports, the harbours are small and narrow, squeezed in between the water and the cliffs. This sea, therefore, is hardly suitable for modern economic developments, which depend more on mass production than on diversity.

We have thus traced on the map several of the principal regions of the globe. First we have the zone of the alpine folding of which the Mediterranean is only one section, and immediately to the north a Primary zone, Hercynian Europe, which is affected by the Alps but does not belong to them. Then the beginning of the immense Euro-Asiatic platform, which includes the great plains of Germany and Russia. Finally, to the south the foundation of the Sahara Desert, a simple part of the Primary continent with no irruptive folding. Suess calls this the Gondwana Block, and includes both India and Arabia in it. The individuality of the Mediterranean region is marked by its opposition to the massive character of Africa and Asia, of which north-east Europe is only a part. From the geological point of view the northern plains lie outside the European region properly speaking, but French North Africa and the Atlas Mountains belong to it. Africa really begins only on the threshold of the desert.

It was thus during the Primary, Secondary and above all in the Tertiary periods that the geological personality of the Mediterranean was formed. From then on there were only tardy or indirect reactions, such as the influence of glaciers, volcanic phenomena, and subsequent irruption of the earth's surface. (Examples of this are the central massif, the Italian volcanoes, the Carpathian Mountains, and the highlands of Turkey and Morocco . . .) Finally there was the influence of erosion in the chalky districts, which was further accentuated by the nature of the climate. Good agricultural districts, which are simply cleavages filled in with soil during the Tertiary era, are rare enough in this harsh broken country.

III

Finally, I must emphasise a geological circumstance which is dependent upon eastern Africa and western Asia, but without which the recent destiny of the Mediterranean could not have been what it is. The continent of Africa is compact and closed, with neither indentations nor peninsulas, a monotonous block of early geological origin. Its isolation from Asia is only apparent, in fact it has not always been so, for Arabia in spite of the Red Sea is only the continuation of the desert of Sahara. Nevertheless it is in this part of the globe that we find more fractures than anywhere else. From south to north a series of linear cleavages constitutes a formidable series of trenches

GEOLOGICAL ORIGINS

extending from the Zambezi to Syria. The southern end of this fissure has been filled in with water, and now forms the great African lakes. The trench corresponding to the Red Sea, now filled with water from the Indian Ocean, terminates to the north like a fork. One prong, the narrow Gulf of Suez, ends with the low lying isthmus, while the other prong, extending through the Dead Sea and the Jordan Valley, continues as far as the Bekaa Valley and the Orontes River, to die out at the foot of the Amanus Mountains. There this enormous geological phenomenon finally ends.

Without this rent in the earth's crust the Suez Canal would have been impossible, even inconceivable. Instead of being a splendid corridor of maritime communication between the continents, the Mediterranean would simply have been an exceptionally beautiful lake. So once again geology has shaped the characteristics which determine the destiny of continents. In the sense of latitude the mere existence of a Tethys would not have been sufficient, for at its eastern end the natural structure was destined to influence circulation north and south. Only in this way could the maritime route to India and the Far East have been achieved.

IV

Thus the Mediterranean has existed ever since the end of the Primary era. At the end of the Secondary era it was merely a narrow trench almost entirely filled in with rubble. Various hypotheses have been put forward to explain how it later reappeared in conjunction with the Alpine folding.

The traditional explanation, which has now become a classic, is that the reliefs, both highlands and lowlands, created very deep fissures. These fissures were the result of vertical movements completing the work of horizontal efforts, which were the natural effect of the Alpine foldings. In these conditions the birth of a sea, complementary to a mountain chain, would be an essential phenomenon of the Tertiary era.

THE MEDITERRANEAN

A more recent hypothesis, originated by Wegener and now adopted by several of the most authoritative geologists, attributes the formation of the Mediterranean to the drift of the continents. In the carboniferous age all the present continental masses may have been welded together in a single block by means of siliceous aluminium, thus forming the solid part of the earth's crust to a depth of say fifty or sixty miles. Beneath this was a viscous foundation on which pieces of the original continents slid in their slow drift. These pieces separated one from another at about the Pliocene era to form the continents of the present day. It was thus that the Atlantic concavity of the African continent counterbalanced the convexity of the continent of South America. The Balkan promontory when it was torn from its base would in the same way have brought the Ionian Sea into being, and then the eastern and western Mediterranean.

The formation of a sea at this juncture would accordingly have been due to the separation of the two blocks of Europe and Africa. This was a fundamental phenomenon corresponding to other separations, and it led to the shape of the present world with its oceans and continents. Once again the Mediterranean seems to be something more than a mere local accident.

CHAPTER II

THE MEDITERRANEAN AS A SEA

THE Mediterranean extends over 940,000 square miles, and if we add the Black Sea we have an area of over a million square miles. If we include all the countries with a Mediterranean climate, this figure is about doubled. Compared with the 30 million square miles of the Atlantic, the 60 millions of the Pacific, the $8\frac{1}{4}$ millions of Russia or the $3\frac{1}{2}$ millions of the United States, the proportions of the Mediterranean obviously are not on the same scale at all. Now the measurements of the civilisation which we inherited from the Greeks are based on the size of this sea, where man is not out of proportion. 'The saying of Protagoras,' wrote M. Paul Valéry, 'that man is the measure of things, is essentially characteristic of the Mediterranean.'[1] The modern world, as it turns on the axis of the new continents, is adapting itself to other proportions. Perhaps this may account for the crisis through which Europe is now passing.

The Mediterranean is a very deep sea, in reality it consists of a string of submarine valleys of which four are nearly 1000 feet deep. One lies between the Balearic Islands and Sardinia, another between Sardinia and Calabria, another between Sicily and Cyrenaica, and finally one between Crete, Cyprus and Egypt. These valleys seem to be extended into four others which are between 600 and 900 feet deep. To these must be added two outlets, one towards the Straits of Gibraltar, and the other flowing over a rise in the sea floor towards the Adriatic. In spite of the fact that the Mediterranean zone is remarkably uniform otherwise, there is thus no unity under the sea, and this emphasises the importance of the submarine highlands which are the remains of previous connecting links.

[1] PAUL VALÉRY, *Variété III*, p. 256.

THE MEDITERRANEAN

The highland of the Balearic Islands attaches that archipelago to Spain, after which deep vertical cañons appear towards the east. One side of the pedestal of the massif of Corsica and Sardinia is linked up with Italy, and as late as the Quaternary era it was still united with the continent. From the coast of Tuscany one can see the Island of Elba with Corsica rising behind it. The pedestal of Sicily and Tunisia is still more interesting, as it reveals that in the Quaternary era there was a connecting link between the two continents. The straits are so shallow that if the waters were to subside another 300 feet there would be only a narrow threshold left beneath them. Perhaps this explains the significance of the traces of dwarf elephants in the soil of Sicily. The Aegean platform lies underneath the Aegean Sea and accounts for the innumerable islands which are always in sight as one passes by boat or aeroplane from Greece to Asia Minor.

One guesses that even beneath these islands, although less clearly defined than in the platforms, the Alpine chains still thrust forward, so that although we lose sight of it under water the general design of the structure appears to carry on without interruption. The great Alpine chain plunges beneath the sea at Nice, turns at the massif of Corsica and Sardinia towards the east, then deflects towards the Balearic Islands. It reappears in two parallel ridges in the Betic Mountains in Spain, and in the folding of the Rif Mountains in Africa. In Tuscany one of the Apennine chains plunges under the water towards the Tyrrhenian Sea, while another ridge goes towards the Gulf of Taranto, after which all trace of it is lost. The folding of the Atlas Mountains running east and west eventually disappears in the Atlantic. The various axes can thus be traced quite accurately; the main axis runs through the Cordilleras, the Betic Mountains, the Balearic Islands and the Alps at Nice, continuing through the Dinaric Alps towards Anatolia, while lesser axes are represented in the Apennines and the Carpathians.

It is not so long since we thought, with admirable simplicity, that we could surround the entire Mediterranean with concentric circles designating the structure of the country. Nevertheless a clearly defined framework does result from these characteristics. Termier actually was well informed when he wrote: 'The Tertiary chain does not run around the western Mediterranean, but actually crosses under the sea at this point. The sea bed here is to a great extent the result of recent sinkings; and if we could descend to the bottom of the gulfs, or better still if we could empty them dry, their walls would appear to be the actual continuation of the folds and tablelands of this chain. Engulfed in the basin of the western Mediterranean we should find the whole of the Tertiary chain, and even wide extensions of this chain.'[1]

Let us now consider the gateways of the Mediterranean, not as an inland sea but rather as an annex of the Atlantic and a highway between the oceans. Apart from the Bosphorus, one of these outlets is natural, but the other — although not less important economically — is artificial.

We know that the Straits of Gibraltar have been open only since the late Pliocene era. Before that time there were other sea lanes, possibly a strait through the northern Betic Mountains, or possibly through the Rif, for this region resembled an archipelago at certain periods. The present strait is only 10 miles wide and 1300 feet deep. After all it is merely a geological accident, although the general direction is accentuated by the parallel folds of the Betic and Rif Mountains. The dominating impression is the unity of the two coasts.

'From the summit of the Sierra Bermeja,' writes Jules Sion, 'we look down upon the white rocks thrown out at the end of the furthest point of the ndy sashore. The Algeciras Mountains seem as confused as a flock of sheep, and beyond them we discern the straits, the sandy shore and the summits of the Andieras. Europe and Africa still seem to be joined together. One recalls a remark made by Suess that the Straits of Gibraltar were a

[1] A. TERMIER, 'Problems of Mediterranean Geology', *La Gloire de la Terre*, p. 94.

simple accident in the architecture of the Mediterranean by comparison with its earlier structure.'[1]

Geologically this opening is not of primary importance. Geographically and commercially, however, it is decisive; indeed the mere fact that there should be a natural passage here is of infinite consequence. One cannot pass this famous landmark without a feeling of solemnity. M. Paul Morand writes forcibly on the subject: 'There is not the same marked opposition between Africa and Europe as there is between Europe and Asia. The mountains dive under water at the columns of Hercules to rise again on the other bank. Looking north and south the view of the straits is not so moving as looking east and west, for then we pass from the Mediterranean into the Atlantic through the great gateway of the western World.'[2] This spot, to borrow an expression from Barrès, is of spiritual significance. To those who are preoccupied with the control of the seas, it also stands for a great deal.

The other gateway, that of Suez, is artificial only in appearance, for everything suggests that a natural communication once existed between the Mediterranean and the Red Sea. An ancient tradition among the Egyptian priests, retold by Herodotus, seems to revive a memory from the dim past. The alluvium of the Nile gradually filled up the strait, while a barrier was forming towards the south, cutting off the Bitter Lakes. We know that the Egyptians made a canal from Bubastes (Zagazig) to Clysma (Suez), but it played purely a local role.

The Mediterranean basin, deserted by international trade since the time of Vasco da Gama, was turned into a sea in the nineteenth century by that great Frenchman, Ferdinand de Lesseps — one must give him his title — bringing the oceans into communication, and thereby releasing a series of marvellous consequences. The economic gain is certain, but politically one is not so sure. Mehemet Ali, the great Viceroy of Egypt, said he did not wish to see his country become a second Bos-

---

[1] ARMAND COLIN, *Géographie Universelle*, vol. VII, p. 10.
[2] PAUL MORAND, *Méditerranée, Mer des Surprises*, p. 133.

THE MEDITERRANEAN AS A SEA

phorus. When Renan welcomed Ferdinand de Lesseps to the French Academy, as we said earlier, it was with these singularly prophetic words: 'You have thus indicated the position of the great battles of the future.'

11

The life of the Mediterranean zone is closely determined by the very special character of this sea. As we know it has no tides, or none to speak of, for the water level never rises more than three feet. This immobility makes people who are born on the Atlantic coast feel that this just is not a sea at all. During the summer the beaches have to be swept like a floor for the water never goes beyond its allotted domain. Boats, however, can enter the harbours at any hour, an important point in this age of speed. The waves are short, hollow and choppy, with a broken rhythm which does not in any way resemble the slow almost majestic rhythm of the ocean swell. On an average the waves are nine or ten feet high, although on occasion they do rise as high as forty. One wave follows another at a distance of fifty or sixty yards. One can soon tell when one has passed Gibraltar for the waves then are a different shape.

There is little resemblance between the Mediterranean and the Atlantic, except during a storm with the wind driving from the east. Owing to the evaporation of the water under the burning sun, the saline concentration here is accentuated. On the other hand the waters of the Black Sea and the Adriatic are less salty owing to the fact that they receive fresh water from the Danube and the Po. Without this inflow and other currents from neighbouring seas, the level of the Mediterranean would tend to subside. One current that is less salty comes in past Gibraltar; another coming from the Black Sea pours like a rapid river through the Bosphorus, and in the same way the Adriatic empties without hindrance into the Mediterranean.

Being thus isolated from the oceans, the marine life of the Mediterranean is poor, all the more so as we go eastwards towards the deepest parts. Nevertheless the Mediterranean fauna has a character of its own. It consists of tunny fish, and such fruits of the sea as sponges, molluscs and coral which have to be specially handled. There are no great fisheries, but the ponds are teeming with life. After all this is a sea not a lake — and yet it is not an ocean. Contemplating this splendid sea, this divine sea — the terms are not excessive — one still longs for the real sea, for something certainly is lacking here. Perhaps this Mediterranean, as one gathers from its name, is too closely surrounded by land. Sailors, however, indeed all sailors, regard it as the cradle of navigation, almost as something sacred. One should re-read the way Conrad describes the Mediterranean in his *Mirror of the Sea*: 'Cradle of the navigation of the high seas and of the art of naval warfare, the Mediterranean tenderly touches the susceptibilities of every sailor. He looks upon it as a vast nursery in an old, a very old dwelling, where countless generations of his own kind have learned to walk. The Mediterranean has cradled the profession in its infancy.'

III

Along the coast we again encounter the direct effect of the geological structure in the contact of the mountains and the great depths to which the sea descends beneath them. The torrential rivers and the absence of tides also contribute to the general plan.

The first consequence is that the coastline is abrupt, almost chiselled out of the rocks, and therefore there is no room for the usual alluvial plains. This structure is seen in the *cales* of the Balearic Isles, the *calanques* of Provence and the *canali* of the Adriatic. When the exposure to the sun is favourable we have a typical Mediterranean coast or riviera. 'Certain coasts,' writes Vidal de la Blache, 'in formation and slope are not unlike fruit trees on a wall, and on these man had only to build his terraces... The mountains rise up close to the shore, we might almost say they envelop it. On the slopes facing the sea we find the principal town, with its whitewashed houses perched between the plantations and the olive groves, and linked with the beach by donkeys toiling up the zigzag path. An excellent example of this is the Ligurian district, which is popularly known by the characteristic name of *Rivière* . . . This type of *Rivière* is repeated the length of the Mediterranean though usually on a lesser scale.'[1]

Along this clear-cut coastline, with its marvellously pure water, the problem of silting, so troublesome elsewhere, does not occur. It is the sheer depth of the sea that poses the problem when it comes to building jetties and piers.

The second consequence of the geological structure is that there are a great many deltas. As the rivers rush down the slopes, they carry with them a vast amount of soil and there is no tide to bear it away, although underwater currents do come to determine its limits or perhaps to form it into plains further along the coast. The ports that are most favourably situated, if they happen to be near the mouth of a river, are those that are sheltered from these currents — Alexandria, for example, in relation to the Nile. Port Said on the other hand has had to be protected from the mud of the delta — there was no choice when it was built — by a specially constructed mole resembling a breakwater. Although this mole has been extended further and further out to sea, the problem has never been completely solved.

The third consequence is that there are few plains in the Mediterranean region, and any that do exist naturally are uninviting. The water precipitated by the torrential rivers accumulates in these plains in stagnant pools or quicksands, and thus creates districts that are unhealthy. But once they have been drained and made healthy, or laid out in irrigation canals, these scenes of sadness and desolation can become excellent and profitable in every way. Once again we repeat, in the Mediterranean man must never relax his efforts to support nature, to correct her and keep her in check. The result still is precarious,

[1] VIDAL DE LA BLACHE, *Principes de Géographie Humaine*, p. 85.

THE MEDITERRANEAN

for the difficult conquests of civilisation always threaten to return to the desert and barbarism.

Finally one last consequence, the Mediterranean coast although it possesses the undoubted advantage of bays and inlets, often also suffers from the disadvantage of being separated from the interior by a barrier of mountains. Thus there are usually no roads leading inland, and above all no navigable rivers. There is no space behind the ports, and even in the ports there is not enough room for warehouses and docks. One economic consequence that is both far-reaching and serious is that the general direction of the trade routes is affected. To serve the needs of European transportation, the routes should penetrate straight inland from the coast. But as it is they have to go the longest way round, either towards Gibraltar or the Black Sea, in order to reach the centre of the continent. The competition of other transit routes is one of the most serious problems that the Mediterranean has to face. This also is written on its geographical structure.

CHAPTER III

THE MEDITERRANEAN CLIMATE

THE Mediterranean climate is so characteristic that geographers have taken it as a type. We have a Mediterranean climate when a desert or a torrid zone occurs alongside a temperate ocean. The contact, the contrast, the double reaction of the desert and the ocean generate a certain kind of atmosphere and vegetation. These in their turn create definite standards, a type of trade and commerce, even a certain way of living. There is even a Mediterranean civilisation which is the direct result of the climate. For example in California we have the reaction of the Rocky Mountain deserts and the Pacific Ocean; in Chili the dry plains of the Andes and the Pacific; in South Africa, high tablelands and the Atlantic; and in south-west Australia, the Australian steppes and the Pacific. Each of these regions has a Mediterranean climate.

Caught between the double attraction of the tropical zone of the Sahara and the temperate zone of the Atlantic, the Mediterranean is dominated during the summer by the influence of the desert. A Mediterranean summer, with its chronic drought and really tropical heat, is not a European summer. During the winter on the other hand, the low pressure of the Atlantic extends over the Mediterranean, and the weather is ruled by the atmosphere of the ocean. Mediterranean winters, with their winter and mid-season rain, are European winters. Thus, by tropical anti-cyclones and Atlantic cyclones, the desert and the ocean make their presence felt by turns. In the summer the whole region becomes African and seems to have no contact whatever with the West, but in the winter it Westernises itself again. As a result of this double participation, the Mediterranean is part of the West; yet, being also African in a way, it is no stranger to the East. This intermediary position brings many valuable advantages, but also certain dangers.

Naturally the extent of these influences varies, according to whether the ocean or the desert is nearby or far away. In winter the Atlantic cyclones affect the entire Mediterranean, but in summer their influence scarcely goes beyond the western Mediterranean or the Adriatic. The course of the cyclones varies in the same way. In winter they pass lower down, from Gibraltar towards the Syrian coast, but in summer they take a more northerly route, making contact with the Mediterranean in the Gulf of Lyons.

Although these extraneous influences may penetrate deeply, the domain of the Mediterranean climate proper still is not very extensive. However, the structure of the framework is now apparent. As the mountains are very close to the shore, the temperate effect of the sea suddenly ceases. In the eastern Mediterranean especially, we find the continent affecting the climate of the sea. Thus we have a striking contrast between the way that the Atlantic influences extend over the European continent towards the East, and the narrowness of the zone of the Mediterranean climate.

II

The general characteristics of the Mediterranean climate are so well known that it is sufficient simply to allude to them: exceptional luminosity, clear blue skies, bright light, atmosphere so transparent that the naked forms of the mountains stand out as clearly as in a dry point etching. The Mediterraneans love these forms, and prefer them to the 'indefinite outlines of the wooded hills' of which Chateaubriand writes, where the sharpness of the profile is lost.

In spite of the summer mists, the sun rises with amazing ease along these coasts, without the hesitations and uncertainties which attend the coming of day in the far north. Jules Tellier of Havre marvels at these Mediterranean sunrises: 'The East grows light, and it is dawn, and it is day. Nothing was ever swifter, nor with less of the richness of a northern twilight. Not

THE MEDITERRANEAN CLIMATE

for an instant have we had an impression of struggle, of a doubtful issue, nor of any obscure resistance from the things of the night. Everything has taken place simply and neatly. Another day is here, that is all there is to it. It has stepped into its proper place. It would seem that the night does not claim these countries for its own, and that it abandons them for the asking. Yet people who begin the day in this way are bound to lose, and do lose, something of the doubt and depth of poetry.'[1]

Renan, who like Tellier lived on the shores of the Channel, would no doubt have approved of these lines, into which has slipped a certain longing for the Western skies, skies that are 'animated and lively as a crowd'. By comparison with this triumphal Mediterranean awakening, an Oxford morning with its solid breakfast seems extraordinarily heavy! A Greek man of affairs does not breakfast at all. He is perfectly content to go to his office on a cup of coffee.

Though the average temperature is kindly, with little variation between summer and winter, there is a vast difference due to the burning sun between the heat of the day and the cold of the night. Sunset is a critical moment, a danger point after the deceptive warmth of the day. It may be that this accounts for the melancholy of the Mediterranean evenings, which poets have evoked and painters have reproduced. I can never read without emotion Maurice Barrès's description of his first contact with the Egyptian coast, on an evening which seemed to be devoid of hope.

'We are nearing Alexandria, which already is tinged with the softness of Egypt. There are long pale lines of extremely delicate colouring, a horizon which has not yet taken shape, an amphibious country with a ghostly Venice floating on a green sea. And while the day closes in with extraordinary sadness, the classical ship of Delacroix, the Levantine bark, is bringing a turbaned pilot out to us.'[2]

By contrast there is little change from one season to another,

---

[1] JULES TELLIER, *Reliques de Toulouse dà Girone*, p. 127.
[2] BARRÈS, *Cahiers*, vol. VI, p. 155.

and only the barest accent on spring and autumn. A European no longer recognises the seasons in this spring which is heralded in at the end of January, or in this joyous autumn which after the mortal drought of summer appears like a resurrection. The sea has no tides, the climate has no seasons, and people from the north are always homesick in these far too happy lands.

The strongest contrasts arise from geographical or topographical circumstances, such as exposure to the sun or the effect of the sea on the mountains. It is usually not necessary to go very far inland to meet the continental climate, which is already apparent on the plains and which soon dominates the mountain regions. To the south, in any case, it is the heat that matters. In January a line drawn at 10 degrees centigrade divides the Mediterranean into two equal climatic sections: in Africa the more southerly significantly includes the cape of the three peninsulas, which lies beyond the atmosphere of Europe.

Drought, almost the aridity of the Sahara, is the outstanding characteristic of the hot season of the Mediterranean climate. It is almost as if the influence of the desert had extended over the sea as far as Toulouse into an atmosphere that otherwise is oceanic. Again the rain of the West is in strong contrast to the rain of the Mediterranean. In the West it always seems to be wet, and yet very little rain actually comes down. On the Mediterranean, however, although rainstorms are rare enough — the average is less than 100 rainy days a year — when they do come they are violent. They release enormous reservoirs of rain in a few hours, and destroy everything in their path. In 83 days of such downpours, Montpellier receives as much rain as Lille does in 203. A single day's rainfall in southern France is equal to that of a whole month in Paris.

Instead of doing good, these downpours do harm. They are destructive, almost catastrophic, uprooting and carrying away the soil to accumulate uselessly at the bottom of stagnant pools. Consequently ceaseless planning and supervision are required to conserve the rain water and direct it to suitable places, and also to drain off any harmful excess and preserve the arable

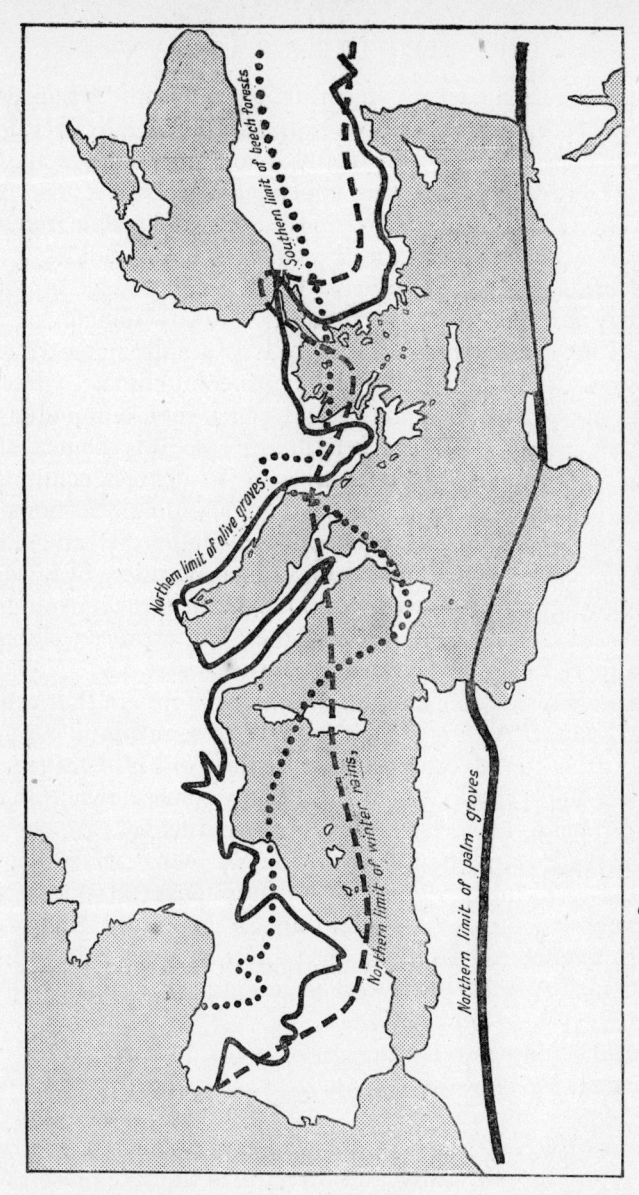

FIG. 8. NATURAL BOUNDARIES OF THE MEDITERRANEAN REGION

land. Out of this an entire civilisation has developed, based on the patient individual effort and co-operation of those whose interests are at stake. Nothing has ever impressed me more than this endless toil, which seems to go back to the earliest days of the human race.

The geographic division of the rainy areas indicates a Mediterranean zone hemmed in by a European and an African zone. Little or no rain falls during the whole year south of a certain line; this marks the beginning of the Mediterranean climate which contrasts sharply with the temperate European climate. South of another line it never rains at all; this with its great palm trees is the beginning of the true African climate. Finally, and still within the Mediterranean climate, there is a region to the north where the rainfall is heavy in the autumn, and another region to the south where it is heavy in the winter. This is the difference between a relatively temperate Mediterranean, and the African Mediterranean which includes the three great peninsulas (Fig. 8, p. 61).

When we study the winds we find extreme conditions again. The mistral and the Adriatic bora, which blow from inland and arise originally from a depression over the sea, are violent, dry and cold, exciting or depressing by turns, but almost always enervating. They do, however, blow away the clouds, and reveal a sky of glorious splendour. But we must pay the price for this glorious splendour, for when the mistral lasts for several days it becomes almost intolerable.

'In a sky swept to purity by an icy mistral,' writes M. Audiberti, 'a burning sun beats down on a cold wind, like enamel on iron ...' This description is excellent, but does it hurt![1]

The winds from the sea, such as the marin of our Midi, are damp and mild, bringing with them a sticky unhealthy atmosphere, with the saturation point as high as 95 per cent. 'The marin comes to us from the sea,' writes an author from Narbonne. 'It is wet of course and brings on rain. As a rule it is a

[1] AUDIBERTI, *Comœdia*, 6 Décembre, 1941.

light wind, but when it blows continuously for several days, the air one breathes at Narbonne becomes excessively humid. If the wind dies down the steamy atmosphere becomes heavy, and the humidity with which the air is laden penetrates everywhere. Walls, marble, stone, all are covered with moisture; in the streets the pavements are wet, wood swells, watery vapours condense on the window panes and run down the surface, and meat goes bad in a couple of hours. Men and animals alike are overpowered. One feels miserable, headachey, weak and too inert to make any effort. This air relaxes the whole system.'[1]

Michelet, who always adds a passionate touch to the argument, says it is difficult to heal a cut limb at Narbonne.

The winds that originate in the Sahara bring heat in the same way, but they are different from the marin, for their heat is dry, violent and overpowering. The Egyptian khamsin darkens the sky with a fine sand which penetrates everywhere, causes a sore throat, and gets into the works of your motor car. An aviator caught unawares when this wind blows has to take great risks with his machine. The sirocco eventually crosses the Mediterranean, carrying with it into France a fine, dry, pulverised mud which covers leaves and flowers with a reddish precipitate. In addition it has a weakening effect, which makes the summer hard to bear. It seems to bring a remote touch of the Sahara to whatever regions it manages to reach.

But the marin, when it continues its course beyond the maritime zone, changes curiously in character. It loses its initial humidity, and becomes the dry, hot autan of the Toulouse region. This is the south wind that is so dreaded by the people of Lyons. Dr. Mouriquand of that city declares that he finds that when it blows, his patients are worse. Now the line where the marin changes into the autan indicates a fundamental boundary, that of the Mediterranean climate. Beyond this aerial frontier the trees are different, production is different, values are different, in fact we have another civilisation. We

---

[1] J. DE MARTIN, *Essai sur la topographie physique et médicale de la ville de Narbonne.*

THE MEDITERRANEAN

have linked up again with the Atlantic. The line is clear cut a little to the west of Carcassone, so definite indeed that one might trace it out on a map almost to within a few miles.

III

This so-called Mediterranean climate exists chiefly along the coast, except when the desert stretches right to the sea. We find it on the maritime slopes of the mountains of the Riviera, in the plains that lie close to the sea as in Languedoc, and when it winds for miles up certain valleys, such as the Aude, the Rhône and the Durance. There are also other climates akin to the Mediterranean wherever the temperate north or the deserts of the south encroach into this region. So summer is not dry everywhere, and some places may remain like the Sahara even in winter.

As a rule, however, the climatic boundary lines are definitely marked. In France one comes upon these frontiers on the mountain-side at a height of 2000 to 2500 feet, where the trees change from chestnut to oak at about cloud level. South of the Mediterranean, on the other hand, this boundary is not reached below 3500 to 4500 feet. Beyond and surrounding the Mediterranean climate proper is a mountainous climate which is essentially due to the favourable lie of the land. The slopes of the Cevennes to the south, for example, seem to enjoy an extension of the Mediterranean climate. When the line is not marked by mountains it may be indicated by the desert, which on the Libyan coast does not leave so much as a fringe of country which could be called Mediterranean. Sometimes a very narrow strip, not more than a few miles perhaps, does represent the Mediterranean domain. Such is the case at Alexandria, where a taxi will take you in a quarter of an hour either into the desert or on to the delta. When there are neither mountains nor deserts, the frontier may be simply where the winds meet! The prevailing wind may lead to all sorts of consequences, even exerting an influence on politics.

THE MEDITERRANEAN CLIMATE

In spite of these variations, the uniformity of the climate is one of the most impressive characteristics of the regions surrounding this sea; it may even be the key to the individuality of the Mediterranean. When we leave this remarkable climate everything changes, from the sky to the psychology of the people. Entering this region the effect is as clear cut as opening a door into another world. A Frenchman no longer feels he is in France, a European no longer feels he is in Europe.

Above all the behaviour of the people is different. This regime, with its contrasts of pressures and winds, of temperatures and sun rays, seems to have a direct effect on the temperament and general balance of the people. According to Dr. Carrel, these abrupt variations increase the vitality. Vitality is in fact one of the characteristics of *homo Mediterraneus*. He is quick and alert; his reflexes according to Daudet are as quick as the trigger of a modern gun. He is emotional, expressive and yet cold at heart. He gives way less to sentiment than the northerner does; he is less practical perhaps but in the last analysis more realistic. At the same time his efforts are often cut short and incomplete.

Contrast this with the patient work carried on along the mountain slopes close to the coast, work which people from the north always find so amazing. But if one lives in these surroundings one must accept these anomalies, which no doubt are the result of a climate which at first sight seems to be too uniformly pleasant and benevolent. The Mediterranean does not exactly live up to its reputation, however. 'Thanks to the Greeks this is a literary sea,' writes Paul Morand, 'so our professors tell us that it is dominated by man, and that it is the seat of wisdom and poetry. This is not true. On the contrary everything that touches the Mediterranean partakes of eruption, torrent and tornado....'[1]

The author of these lines knows the Mediterranean so well that one must believe what he says. The publicity of the Midi written to attract the winter visitor is couched in entirely

[1] PAUL MORAND, *Méditerranée, Mer des Surprises*, p. 5.

different terms, and yet it is not wrong either. Perhaps the truth may lie, as Renan suggests, in a dialogue? What an interesting geographical essay one could write by combining two different avenues of thought: (1) A eulogy praising the climate of Nice as drawn up by the Association of Hotel Proprietors; (2) A summary of the climate of Nice written for a meteorological dictionary. The truth would lie somewhere between the two, but if I were to ask for such an essay some suspicious Cato would certainly hound me out of the Republic! Oh well, Daudet never tried to unite his Tartarin-Quichote and his Tartarin-Sancho. . . .

CHAPTER IV

THE FLORA OF THE MEDITERRANEAN

THREE distinct circumstances determine the character of the vegetation of the Mediterranean.

First of all, the effect of the climate. This whole region is dry, with a dazzling sky, hot summers and mild winters, yet cold enough for plants to feel the difference. Consequently they must rest during the winter. But they must also rest in the summer, especially in Africa, if they are to survive the prolonged droughts of the hot season.

The structure of the country also exerts an influence. The mountains are close to the sea, and the coast is indented with little bays which divide it up into sheltered compartments. The sea, being always close at hand, has an equalising effect, cooling the air by day and warming it up again at night. In addition the mountain barrier, in Europe at any rate, protects the vegetation against the north wind, while the sea breezes tone down and correct the excessively cold nights, especially after hot days. Therefore exposure to the sun is very important, all the more so as it is a southern sun in a country which basically is cold owing to its latitude. This accounts for our misunderstanding of the Mediterranean.

'Is the weather really good on this coast, this damned coast?' asks M. Audiberti, 'or is it bad? The temperature is always skipping about. Between the sun's burning rays, air currents cut in that are as hard as the edge of a scythe. The luxurious appeal of the landscape in no way deters the deplorable effects of arthritis. A glimpse of Paradise goes hand in hand with the malediction of neurasthenia. The spirit hesitates between the illusion of perpetual summer and the temptation to freeze.'[1]

Finally, there is a whole series of circumstances which arise

[1] AUDIBERTI, *Comœdia*, 6 Décembre, 1941.

THE MEDITERRANEAN

from the nature of the soil and the water supply. The soil generally is calcareous, very hard, with steep slopes on which the climate scarcely permits vegetation. The rivers are swift, spasmodic, and liable to terrible erosions. As a result cultivation is precarious, and requires unremitting care. Good soil is to be found in some districts certainly, but it is localised and relatively rare.

These conditions, as we see, are at once exceptionally favourable and exceptionally discouraging. A marvellous growth is possible, but always with the threat of total extinction or of returning to the desert pure and simple. It is magnificent, but it is precarious.

II

The effect of the seasons then comes to complicate this first set of circumstances. Oddly enough the heat of the summer has less influence than the absence of cold during the winter. Two essential factors of the climate, heat and humidity, are out of harmony.

In point of fact heat seldom coincides with humidity. In summer there is practically no rain, or at least not enough to affect vegetation. In autumn and spring the downpours of rain, erratic and tumultuous, are almost useless, for the sun's rays and the mistral quickly dry off the soil before it has been watered enough. However, there is still a certain amount of irrigation, for the rainfall is better distributed on the mountains which are always close at hand, and they act as a reservoir. In these conditions, as heat and humidity come along separately, either the plants must find their own supply instinctively, or else man must come to their aid artificially. Sometimes heat and humidity do coincide. Then the result is excellent, and one is almost tempted to forget the very real handicaps of the Mediterranean region.

The rhythm of the seasons thus lays its imprint on the vegetation. Plant life has two periods of rest: a short one

THE FLORA OF THE MEDITERRANEAN

during the winter, and a long one (especially in Africa) during the summer. They also have two periods of activity: a quick burst into life in the spring and even earlier which dries up at the first sign of heat, and a sort of autumnal spring when the first rains begin.

Thus the autumn gives the impression of spring — luminous beauty, and a sensation of rebirth after the first rains during the purified days of October. But this is simply an appearance of rejuvenation, for the melancholy of the shortening days is there just the same, exerting its subtle presence even in this climate. It is in January and February that one must look for the real spring, which comes with an outburst of dynamic joyful light, and days which express the eternal and blinding youth of Nature. As for the summer, it blossoms forth at the moment when the true spring — so capricious in northern countries — begins to establish itself. In the long months of drought, everything goes to sleep. Plants grow in the autumn, rest in the winter, flower in the spring. Then everything stops, and they rest again during the summer. Such in a general way is the regime of the Mediterranean. The exotic vegetation which flourishes all year round disturbs this order of things. It is not quite typical of the Mediterranean, and the illustrated postcards certainly make too much of it.

In the Mediterranean region the rocks are always ready to tear a hole in the light top soil. This must be set down, as we have seen, to the type of soil and to the climate, but man, who by nature is destructive, also is responsible. Very good earth like black loam is the exception, and so is the hard granite soil which is often found on tablelands which do not drain off easily. Most of the soil is calcareous, but there is also a great deal of salty land in the marshes where the water has not drained away. This may account for the steppes which are typically Mediterranean and are to be found in Spain, in North Africa, in Macedonia and in Asia Minor. The main thing to remember is the distinction between the thin bright granite soil of the *garrigue*, and the flinty soil of the *maquis*.

THE MEDITERRANEAN

Owing to the hard type of rock which is just below the surface, the soil generally is poor, and such vegetation as exists is constantly threatened. This country could never be used for pasture, nor for cultivation on any great scale. And yet whenever there is any soil it is excellent, and if in addition there is any rain and the situation is good the result can be almost incredible. But it is always at the price of effort, almost of construction, and whatever has been conquered still has to be protected.

Plants do adapt themselves to these complex and special conditions, however. Their principal reaction consists of a system of slow but steady assimilation. As the favourable periods are too short to permit a regular caducous foliation, they seem to find it best to restrict the power of assimilation to their foliage, and remain active in all four seasons. The plant structure consequently is modified, for the chlorophil surface is reduced. The leaves are small, inclined to be leathery and are hairy on the reverse side; often they are rolled rather than flat, and grow parallel to the rays of the sun. In some cases they almost disappear. Thorns are frequent on shrubs and undergrowth, and seeds mature rapidly. Reserves of moisture accumulate either above ground in thick-leaved plants of the cactus type, or below ground in plants of the bulbous, rhizome or root-stock type. Below the surface the roots tend to penetrate very deeply in their resistance to drought in order to find moisture somewhere.

One gathers that in these conditions trees with caducous leaves are not in their natural environment. They do, however, flourish in the damp valleys where one finds the splendid plane trees that Van Gogh used to paint, or on the mountain slopes where we have the fine chestnut trees of the Cevennes and of Corsica. This territory is more suitable for ligneous or woody plants, which are always green and have small leaves and many thorns.

Finally, we must mention the spring vegetation, which flowers suddenly everywhere, even among the most inhos-

THE FLORA OF THE MEDITERRANEAN

pitable rocks. That again is typically Mediterranean, more indeed than the thick-leaved plants which are associated with the Côte d'Azur and the African beaches, although many of them, such as the fig tree and the aloe, were originally imported from Mexico.

The scenery of the Mediterranean is the result to a great extent of these special kinds of vegetation. The severe colouring of the Midi belongs to the dry tropical vegetation, from which the rich greens are absent, and to it we also owe the splendour and scents of the mountainside. On alighting from the train at Marseilles the traveller, whose lungs are still filled with the fogs of the north, is enchanted to be greeted with such an exhalation of perfume.

III

This vegetation extends along two strata: first, a fringe along the sea coast which constitutes the Mediterranean zone proper, and then along its extension up the maritime side of the mountains; secondly, the mountains, or in some cases the desert, which provide a double framework closely attached to the picture but yet remaining outside it. When we arrive at the mountains or enter the desert, we definitely have left the Mediterranean region.

The flora of the coastal fringe comprises mainly the forest, shrubs and clusters of evergreens. The shrub is typical of the first stage. Through the dry soil it reaches the humid subsoil without irrigation, thanks to its deep roots. This opens up exceptional possibilities for the cultivation of shrubs in these districts.

The olive is the classical example of the Mediterranean type of tree, and its limits coincide exactly with the limits of the Mediterranean zone itself (Fig. 8, p. 61). Thanks to its long roots, it can resist the drought successfully, and also the furious attacks of the wind. 'A great root has emerged from the earth to live in the sky', says Paul Morand. In France the olive tree dominates the scene up to a height of about 1200 feet, and even higher when the situation is favourable. It also insinuates itself far up the valleys, bringing with it something of the atmosphere of the Mediterranean. In Sicily it grows as high as 4000 feet. In Africa it stops where the great palms begin. Grape vines, arbutus, laurel, cypress and citrus fruits, each in their own way, make the most of the subterranean moisture, and all are indigenous to the region.

The Mediterranean forest has its own particular characteristics. It is poor as the result of the barrenness of the soil; the trees are small, the leaves are thick and scarce, and there is little dense undergrowth. As these woods are surrounded everywhere by men and their herds, they are the prey of fires every summer, so are reduced to a minimum in the end. Sometimes, as in the case of the cedars of Lebanon, only a few stray trees survive.

Paul Morand refers to 'these miserable little twisted evergreens of the Midi, these famished forests which can never satisfy their hunger, and yet give back their poor nourishment in balm and perfume'.[1]

The timber is divided into five main types. The evergreen oak, which grows everywhere on the sea-coast alongside the pine, is also found higher up, even higher than the olive trees. In favourable situations it traces out a sort of extension of the Mediterranean zone. It is irregular, bushy, and with its little grey leaves is a definite part of the Mediterranean landscape. The cork tree or cork oak which flourishes on hard granite soil climbs as high as 2000 to 3000 feet in Morocco and Corsica, and up to 3500 feet in Sardinia. The parasol or stone pine grows along the sandy beaches and in limestone valleys, forming magnificent forests on the coast of Tuscany, for example. The Alep pine, along the hot low-lying coastal stretches and on certain poor calcareous lands, coincides, at least in France, with a super-Mediterranean zone. Here it provides an effective note in the scenery with its light green needles and silver bark. The

[1] PAUL MORAND, *L'Homme Pressé*, p. 31.

THE FLORA OF THE MEDITERRANEAN

maritime pine, mediocre and twisted, which one finds on sandy beaches, is associated with the mountains of Maurès and Esterel with their granite soil. It also is typical of the Mediterranean scene, with its outstanding height, its bark and scales, and its thick stiff leaves. In Africa we must add to this list the thuya of Barbary, and the scented juniper of the Atlas Mountains.

This forest is quite different from either the forests of the north or of the tropics, both of which are richer and more densely wooded. The feeling of these southern countries is not so much of forest, but rather of *garrigue* and *maquis*.

The clusters of scrub on the moorland represent a lower stage of vegetation, or shall I say land claimed from the desert at the limit of total drought. By means of these shrubs the southerner, who may still be in touch with densely populated cities, remains also in immediate contact with virgin nature, with a life free and untrammelled by civilisation.

The *maquis* holds sway over these hard, flinty lands. No one has ever described them better than de Maupassant: 'The impenetrable bush made up of evergreen oak, arbutus, lentisk, alatern, heather, thyme, laurier and myrtle, is intertwined and mingled like locks of hair. It is laced together with clematis, enormous fronds of bracken, rock-rose, rosemary, lavender and bramble, and then is thrown on the back of the mountain like an inextricable fleece.'

Or this from Jean Aicard which is scarcely less forcible: 'Deep thickets dominated by gorse, so high, so well defended, and so strong when it is old that often it refuses to bend and one is forced to walk in zigzags around it . . . Lost in the middle of this scramble of thorny shrubbery which forms a hollow vault over his head, the bewildered hunter is seized with sudden panic. Around him the flexible bushes press so close that it requires great effort to part them and take a step through them. Then they close in behind him, and when he has parted them once more with great effort, they slip back into place instantly of their own accord. He might be surrounded by intelligent, hostile wills, all leagued together against him. This forest, low and light, has the elasticity of the fluid element of the sea, which gives way before the swimmer without ever ceasing to grasp him with its even pressure.'

Thus, the general aspect is that of a closely knit impenetrable mass, into which man may be absorbed and disappear. In Corsica the saying is, 'Take to the *maquis*', and in Brazil, 'Take to the *sertao*' — in other words flee, take refuge in the wide open spaces of the plains.

The principal shrubs of this wilderness are the giant heather, arbutus, broom, myrtle, lentisk, turpentine, rose and thyme laurel, sauce laurel, and the laurel of Apollo, furze, the Montpellier rock-rose and the cistus with sage leaves, alatern, prickly pear, liana, clematis, sarsaparilla, madder, honeysuckle and wild grape. What poetry lingers in the very names!

The *garrigue* is the product of the limestone wilderness, where it lends its own colour to the light transparent rocks. It is neither so closely knit nor so high as the *maquis*. One comes across both men and animals here, in fact this is a happy hunting ground for the man with a gun. People do not take to the *garrigue* as they take to the *maquis*. Even in Spain or Africa the *garrigue* is only a transition towards the desert or steppe, where there is only the merest trace of vegetation, no trees, no shrubs, only tufts of green pushing through the ground.

The most characteristic tree or shrub of the *garrigue* is the kermes oak, and it can scarcely be called a tree with its knotted trunk and its leathery leaves. It belongs to the same botanical family as many of the other species to be found here, such as gorse, lavender, thyme, rosemary, juniper, lentisk, maple and Montpellier maple, liana, daphne, alatern, phillyreas, fig, pomegranate, cotton cistus, rock-rose and Spanish broom.

South of the Mediterranean the hill country consists of the Mediterranean slopes of the mountain range, in so far as its direction and altitude allow it to benefit by the southern climate. This is an extension, a transition towards less favourable conditions of heat, but also towards a greater degree of humidity and less drought. Here we find growing alongside each other such Mediterranean plants as can survive, as well as the first signs of mountain or rock plants.

Typical of the region is the evergreen oak, which flourishes everywhere, and may even reach the summits of the mountains. On these granite soils alongside the evergreen oak we find the chestnut. The latter is especially suitable to the slopes where it provides splendid forests, with an undergrowth of golden light and shade cut across by stretches of pasture land. In the Cevennes the chestnut survives as far up as 2700 feet. The pubescent oak with its leathery leaves is the forerunner of other climates, but it adapts itself to the blaze of the southern sky. On these foothills the growth of the *garrigues* is sparse, and mingled with plants of the thick-leaved cactus type.

Arriving from the north, when we cross a mountain pass and come out on to the Mediterranean slope, the colouring, the brilliant light, the feeling of relative dryness, all tell us at once that we are looking into another world. No other frontier in Europe has the same significance. These lofty regions form a great frame for the Mediterranean, beginning on the edge of the plains of central France, and continuing across the Alps, the Apennines, the Dinaric Alps, through the Balkans, Asia Minor, the Iberian plateau and the Pyrenees.

The mountains really start when the Mediterranean trees end. At this point we have a different climate — other types of vegetation, and another way of living as there is always a great deal of permanent snow here every winter. There is not much warmth in the summer and insufficient heat from the sun, not to mention unseasonal frosts both late and early, so the periods of growth are very short.

Once again new types of trees signal the passage into another region. The evergreen oak and the chestnut carry on up to cloud level, up to 3000 feet in France, shall we say. After that the oak, the beech and the fir tree take over. A map showing the dividing line between the various species also indicates the real boundary of the Mediterranean. In the south, notably in

Africa and especially in Libya, the line naturally runs at a much higher level.

If we go without stopping from the sunny low-lying countries, we receive a truly pastoral impression of freshness and shade. We should like to rest in meadows so fresh and green, linger in the shade of these great trees, play the flute and tend a flock of goats. But our impression is too hasty. When winter comes we shall find all the harsh ruggedness of the mountain life which is so different from the easy existence of the lands in the sun. Vidal de la Blache marks this contrast in a few telling phrases: 'Farther on we come to the foothills bordering on the Mediterranean. The South seems to the Northerner to be a country of fruits, as in the same general way Central Europe seems to the man from the Mediterranean to be a country of forests.'[1]

The first sight is splendid, an impression of light, sunshine and joy. But one is soon aware of the poverty, of the precariousness of it all. Something seems to be pressing down, devouring and destroying everything. It is the lack of harmony between a victoriously brilliant light and a region which is not quite tropical. The situation and irrigation, whether natural or artificial, make everything possible, but it must be within the framework which has been dictated almost everywhere by the geological structure. There are few open spaces which would lend themselves to rich harvests. This country is diversified and delicate in its formation: that in fact is its essential characteristic.

[1] VIDAL DE LA BLACHE, *Principes de Géographie Humaine*, p. 81.

CHAPTER V

LANDSCAPE AND SCENERY

FIRST let me give you a few quotations touching on various aspects of Mediterranean scenery:

André Gide (*Départ de Marseille*): 'We leave Marseilles in a violent wind that sways the masts, but the air with its early warmth is grand. The sea is glorious. It is feathered with whitecaps, and our vessel is buffeted about on the crest of the waves. What a feeling of power and glory!'[1]

Paul Valéry (*Les Ports Méditerranéens*): 'Our great Claude Lorrain... exalts in noble style the order of the grand Mediterranean ports in their ideal splendour. Genoa, Naples, Marseilles, all are transfigured by their architectural setting. The silhouette of the hills and the glimpses of water together compose a theatrical scene in which a single actor will presently appear, sing and even die. That actor will be the light.'[2]

La Comtesse de Noailles (*Syracuse*):

> Le matin s'éveillait, tempetueux et chaud,
> La mer, que parcourait un vent large et dispos,
> Dansait, ivre de joie et de lumière infuse.
> Sur le port, assailli par les flots aveuglants,
> Des matelots clouaient des tonneaux et des caisses,
> Et le bruit des marteaux montait dans la fournaise,
> Du jour, de tous ces jours, glorieux, vains et lents.[3]

---

[1] ANDRÉ GIDE, *Les Nourritures Terrestres*, p. 149.
[2] PAUL VALÉRY, *Variété III*, p. 247.
[3] COMTESSE DE NOAILLES, *Les Vivants et les Morts*, p. 125:
    The morning dawned, stormy and hot,
    The sea ran under a wind wide and fresh,
    Danced, drunk with joy and flooded with light.
    Right to the harbour, battered blind by waves,
    The sailors nailed down boxes and casks,
    And the noise of hammers came up through the heat
    Of this day and all days, glorious, idle and slow.

THE MEDITERRANEAN

Paul Morand (*La Méditerranée*): 'To me it is a pit surrounded by steps over which the nations are leaning, all crowded together. I can see their heads on the high plains of Auvergne or Thrace, Libya or Aragon, of Russia or Syria. In this Mediterranean amphitheatre the same piece has been played for thousands of years ... As in the ancient tragedies, the stage is of stone roughly hewn out of the rock. The back-drop is the sea, this same sea which in the theatre at San Carlo once served as a background for La Muette de Portici.'[1]

Audiberti (*Antibes*): 'We go up on the ramparts. The stone is burning hot. Before us lies a great liquid expanse, without a sail or a trawl. To the left the mountains, while to the right the Garoupe gathers the horizon into its generous grasp, stretching away wide and clear. I always see the Mediterranean like this, idle and empty, with the pale shape of a checkered sail travelling across it in the far distance.'[2]

Now let me recall my own journey along the Mediterranean, made by air in two stages:

From Marignane to Beirut: At five in the morning after a lovely night the air is laden with the perfumes of Provence, and the airport is still bright with a thousand lights. The aeroplane — a hydroplane — leaves at about daybreak. After taking off from a lagoon, we are soon over Marseilles, which lies clear as a relief map beneath us. It is the same pale yellow colour that I shall presently see again at Athens. The sea is without a ripple, and little fishing boats look like insects tenaciously tracing their furrows. Toulon, the peninsula of Gien and the islands of Hyères lie sketched out as if on a chart, and then we speed away from the coast towards the south-east.

At this moment the whole Alpine chain appears against a cloudless sky. I recognise the familiar profile of the mountains behind Nice with their amazingly jagged peaks, and more to the north snowy masses which I take to be the Viso and the Meije ... The scene is geographic, almost astronomical.

---

[1] PAUL MORAND, *Méditerranée, Mer des Surprises*, p. 234.
[2] AUDIBERTI, *Comœdia*, 6 Décembre, 1941.

LANDSCAPE AND SCENERY

Corsica soon appears out of a cloud of mist. The vigorous spine of its mountain range is fawn-coloured tinged with blue reflections. Ajaccio is sighted, and we then cut across the southern end of the island. Several hours pass with nothing but the blue Mediterranean, and by eleven o'clock we reach Naples, or at least a little airport about twelve miles to the north of the town. The aeroplane alights in a pond in the middle of a marsh. We have breakfast in a little cabin erected on pilework which is not unlike the huts used by duck-hunters. Towards the south we can discern the outline of Ischia, with Vesuvius rising nearby in an atmosphere of exquisite delicacy.

Soon we are off again. Without delay we pass in front of Naples and above Amalfi. We then reach the Apennines, and cross them almost without realising it, they seem so narrow. On the other side, overlooking a great plateau, the design of the map appears clearly below us. We soon pass over the toe of the Italian boot, and cutting across the Gulf of Taranto we are looking down upon the heel. The heel is laid out like a panorama, a vast, tilted, ochre-coloured plain. I count over twenty towns lying white and extraordinarily close to one another. Then the Adriatic, swept by a brisk wind and dotted with white caps. Another hour over the sea and we are at Corfu, which seen from a distance looks like a rock out of the desert.

We stop at Corfu just as the sun is about to set. The part of the island which faces the continent close by is entirely different in feeling to the side which looks out on to the sea. The latter is poetical, rich in colouring, and framed with dark cypresses; there are olive trees with the light shining through them, grape-vines and fruit trees. So far there has been little change. This might be our Mediterranean Midi once more, with just a touch of the exotic about the Greek life of the streets and the pink cube-shaped houses which one sees everywhere in eastern Europe. The difference is not so much in space as in time, for this is still a civilisation of craftsmen and peasants. In the market-place the stalls recall an earlier period, and so do the innumerable little donkeys which were hurrying in from all sides on the following morning when we were leaving Corfu again.

We started at six o'clock, but we were obliged to get up at three, as we had to put our watches forward two hours to eastern European time without the transition of one hour to central European time.

It was still almost dark when we took off, but the sun rose soon afterwards, revealing an amazing landscape of islands and fiords. An hour later we were forging ahead through the middle of the Gulf of Corinth, with the Peloponnesian Islands to the right and the heights of Parnassus to the left. As it is against the rules to fly high, we were able to see everything much as if we were on the deck of a boat. We arrived quickly at New Corinth, which is laid out in squares like an American town, then the canal, cut so deep that we could hardly see the very blue water at the bottom.

We alighted on the roadstead of Athens, with Sigina to the right and Salamis to the left. My heart leaped as I saw Mount Hymettus towards the east, Lycabettus, and finally the Acropolis, where we could pick out the Parthenon quite clearly. Piraeus is the tint of old ivory, as if it had been built of bones that had been dried and polished. The mountains on the horizon are diaphanous, fairy-like, but the blue of the sea on the other hand seems hard and solid. The resemblance to Marseilles is very striking, for the latter has certainly remained a Greek city. Returning some years ago from South America, I could easily have believed I was in Greece when I disembarked from the ship beneath Notre Dame de la Garde.

After a brief interlude for a cup of coffee, we were flying above the Aegean Sea, with its dozens of islands scattered about, in fact we were always in sight of land until we reached Rhodes. Some of these islands are very flat, but all are of the same oval shape, which gives the impression of the vertebrae of an animal emerging from the waves. They are ochre in colour, but tinted with rose. I picked out a few rare trees, but thousands of terraces under cultivation. We flew above Rhodes,

LANDSCAPE AND SCENERY

then a few moments for a hasty lunch at Castel Rosso, one of the Dodecanese Islands that formerly belonged to Italy. This particular island is tiny, and lies quite close to the coast of Asia Minor. The little town with its pink houses all square and huddled together gives for the first time a definite impression of the East. Up to this point the atmosphere had been much the same as at Marseilles. The flight over Cyprus accentuated the impression that we were approaching Asia, for as we crossed the northern part of the island everything looked shockingly barren.

Another hour over the open sea without sighting even an island and we were at Tripoli in Syria. Here we struck a violent storm, and all we could see was black driving rain as we came down on the harbour. From Tripoli we drove by car for an hour and a half to Beirut. There we found another Riviera curiously like our own, indeed if it had not been for the Oriental life of the streets we might have believed we were in France.

From this rapid flight on the wings of a bird I retain two impressions. First, like Gide writing of Marseilles in *Les Nourritures*, I still think of the glory of it all! The blazing light on a sea of azure, and those dream islands which were so sharply defined that they formed a landscape of incomparable beauty. My second impression is the essential unity of the Mediterranean. It is always the same whether one is at Marseilles, at Athens or on the Syrian coast. There is in truth a Mediterranean civilisation, based on the way the different countries bordering on its coasts resemble one another. The real frontier is elsewhere — at the gates of the desert, or at the entrance to the Nile Delta which really is Asiatic. On this trip I travelled nearly 2000 miles, but I never looked beyond this Mediterranean horizon within which no Frenchman can ever feel a stranger.

From Marignane to Alexandria by air: On that particular day the lagoon at Berre is lifeless for the sky is grey. When we reach the sea after having flown over a thin strip of coast as dry as a skeleton, we can see only a dull leaden sheet of water under the low clouds. Then, above this sea of clouds we fly through open sky above an immense mass of cumulus which looks like an enormous glacier. We veer towards the south-east and I keep expecting some summit in Corsica or Sardinia to rise up out of this ocean; but there is nothing to be seen, and another hour passes by. Then suddenly, very far below us, land appears but not from the side that I had expected. It is low-lying country, burnt brown, with lagoons and large villages in which the houses are all clustered together. It was not for some time that I realised that this was Africa, and that we were heading for Tunis ... Tunis with its deep gulf framed with distant mountains, and its large city spread out over the plain.

The French town is uninteresting. The main avenue in spite of its palms might be in any of our provincial towns. The Governor's residence resembles a *préfecture*. The cafés are the same as any of the cafés in the Midi, except that they are filled with people wearing the fez who chat as they smoke just as they do in Cairo. But towards La Goulette and the gulf the scenery is fine, and the landscape achieves a certain grandeur. To the north is Cap Bon overlooking the blue Mediterranean; to the south two mountains like twins preside over the scene with a certain indifference. In the distance the colours are sombre, black and deep violet, but in the foreground the tints are light for it is spring, and the Arab villages are a startling white. We are well into the Orient by now, and everything is quite unlike the Riviera. All here is sad, as if announcing the melancholy of the eastern Mediterranean. Algeria looks towards the west with a presentiment of the Atlantic, but Tunisia looks towards the east with a presentiment of the sadness of Egypt.

It is six o'clock on the morning of the following day, and the sun is rising as we take off. In the twilight the lagoon seems sad. To the south the twin mountains stand out black and clear like Vesuvius seen double. The aeroplane skims across the gulf in a few minutes, and heads towards the east above the beige-coloured mountains of the desert. Once again we are over the sea, but this time it is a great expanse of azure, dazzling with light beneath a cloudless blue sky. We continue on towards the south-east, and for three hours look out upon the glorious Mediterranean, immobile and idle as it has been since the beginning of time. We are flying very high, at least 6000 feet I imagine, above innumerable little white clouds below which the blue water looks hard and solid.

After leaving Tripoli, there is nothing to be seen but a rather tiresome aerodrome where groups of Italian planes are looping the loop in formation. We pass above several orange groves, but otherwise the desert and the sea dominate the scene as far as Benghazi. This little white city is situated on the coast, enclosed within its walls and ramparts, and lying completely still between the solitude of the desert and the solitude of the sea. Not so much as a boat is to be seen; the waves which break in white foam are the only sign of life. Then the desert again, nothing but limitless desert, stretching away, immense. Seen from the aeroplane it seems to be rolling out its violent hues like a richly coloured rug. The background is tawny or red, with rocks as dry as if they were enamelled, standing out brown against everything else. Not a tree in hundreds of miles, only tufts of grass of a charming amethyst tint, or little bunches of yellow flowers like broom. One imagines that some underground river must account for the long tracks of greenish moisture which follow the depressions. Along the shore the rocks seem to rise red against the blue sea, which stretches away marvellously clear without a trace of boats, villages or other signs of life. Here the Egyptian coast recedes somewhat, so to reach Alexandria we must cross the sea once more. For two hours we see nothing but sky, clouds and, very far below, blue waves.

It is now half-past four, and as we are due to arrive at five o'clock everyone is scanning the horizon to the east hoping that something will appear. But still there is nothing, always nothing whatever. Then, without warning, a fairy-like apparition, a dream city lit up by the slanting rays of the setting sun, a rose-tinted city rises out of the waves. Lying between the Mediterranean and Lake Mareotys, I recognise the narrow, long harbour and the city of Alexandria, with its white houses huddled against each other, and the aerodrome towards which we are steering.

By the time I reach the centre of the town by motor car, the sun has almost disappeared, the atmosphere is blue and the street lamps are lighted. Then, as I go into my hotel bedroom, looking out over the old harbour which is shaped like an egg, the stars are already shining in the sky. I feel the dampness rising out of the sea, which is so near that I might be on the deck of a ship.

Even from these scattered notes it should be possible to discern a few general characteristics of the Mediterranean landscape. Like Gide, I retain above all the dominating impression of glory, of brilliant light full of joyous vitality, which knows nothing of the north with its mists, nor yet of the tropics with their overpowering heat and rain pouring from leaden skies. The harmonious lines in the construction of this country are in the measure of man. In this picture there is nothing colossal, nothing exaggerated.

In this classical scene, which is almost too beautiful, we receive an overpowering impression of tremendous antiquity, almost of eternity. There are too many memories here, too many memories of greatness. Contemplating this unchanging, imperturbable splendour, one has a sensation of the vanity of effort, that time stands still. Yet back of it all, and in spite of its glory, there is sadness. 'Glorious, idle and slow' — how well the Comtesse de Noailles described her days on the Mediterranean!

This landscape has so definite a personality that its gradations and limitations are easily apparent. At certain points the zone is so narrow that it can be crossed in a few steps. The short cut proposed by M. Paul Morand in connection with the French Riviera at Beaulieu is striking: 'A few inches of Brazil overlaid on the structure with thirty feet of Greece behind, and already

LANDSCAPE AND SCENERY

we perceive the skeleton of the theatre: the Alps.'[1] This formula would apply equally well to the Italian Riviera, to Sicily, to Greece, to western Asia and to some extent to North Africa. The botanical wealth of the gardens at Cannes and the tropical avenues of Monte Carlo recall the Gloria district of Rio de Janeiro. Marseilles is the sister of Athens, but whether one thinks of Lebanon, the Taurus Mountains or the Alps, it is always the same powerful framework of mountains that encloses these enchanted shores.

Geographers tell us that the olive zone coincides exactly with the Mediterranean zone, and of this there is little doubt. The frontier of the para-Mediterranean zone is more subtle, but the evergreen oak shows pretty well how far it extends. The Mediterranean zone, by a thousand indications, goes far beyond that. In a perfume, in a reflection, in an accent, it travels away beyond its own territory, and although it may be *incognito* one can always recognise it.

At this point arises a series of questions, to which I shall not try to reply. To ask them will be enough. If I come from Paris to Marseilles, where shall I meet the Mediterranean? At the Rove tunnel, naturally, but the Mediterranean will have come to meet me long before that! Where will I feel the first distant approach of the Mediterranean atmosphere? When will I first notice the genuine colours of the Mediterranean?

Now another problem, but of the same kind: if I go from Beirut to Damascus, from Jaffa to Jerusalem, when do I leave the Mediterranean and enter the domain of the desert? And if I go from the Sea of Marmora to the Black Sea by way of the Bosphorus where do I meet, not the Black Sea itself, but the atmosphere of the Black Sea? Finally, if I pass out of the Mediterranean through the Straits of Gibraltar on to the ocean when shall I have the first sensation of being on the Atlantic? The reply to these questions will give us a precise definition of the Mediterranean scene.

[1] PAUL MORAND, *Le Figaro*, 26 Mars, 1936.

THE MEDITERRANEAN

II

The Mediterranean is a blaze of light; actually there is more light than colour. The light is vivid, but the colours are vague, sometimes so faded that they cannot be seen without light. 'In the autumn,' writes M. Paul Morand, 'the Midi is like a tap dancer who has borrowed everything from the sun. Leaving all the violent colours to the northern countries, the Mediterranean is the quiet refuge of delicate tints, of fawns, of rose, soft white, rusty gold, and the silver grey that was as dear to Corot as to Picasso. It is the land of melodious rhythms, not of discordant notes and disjointed tumult. It is the kingdom of style which, though often hidden, orders the mass of rough cyclopean constructions in the same way as infinitely minute mosaics.'[1]

Under these skies, therefore, we must above all love the light. It frees the outlines of the landscape, and emphasises its smallest details. The art dealer, Druet, when he was once showing some pictures to a customer, selected one of the Midi by Marquet. 'Now,' he said quietly, 'you are going to see light.'

Actually we cannot do better than to take the following as guides. In literature read Fromentin and Daudet; in painting follow Claude Lorrain, Delacroix, and above all Renoir, Cezanne and Van Gogh, but also Besnard, Mathis, Marquet, Dunoyer de Segonzac and Othon Friez. In their pictures one recaptures the basic colours which give the atmosphere its individuality, the dazzling white of the limestone rocks, the old ivory of the cities which have outlived the past, the lapis-lazuli blue of the sea, the faded green of the leaves tending to iron grey, the deep blue of the sky and the amethyst background of the sunsets spangled with gold. On our travels we note the finer shadings — in Spain where the ocean is tinged with azure, or on the Bosphorus where the water at sunset is the blue-black of the Styx. When these colours cease and give way to others, then we know we have left the Mediterranean.

[1] PAUL MORAND, *Méditerranée, Mer des Surprises*, p. 24.

LANDSCAPE AND SCENERY

One of my friends, the artist Bérény, has summed up for me the basic colours of the district of Vence in the Alpes-Maritimes. 'On a sunny day', he says, 'the air is blue; the sky blue-green; the light, hot and intense. The distinctive feature is little or no moisture, hence the absence of vivid green and light leafy vegetation. The air contains fine grains of dust. The greys and ochres of the earth will warm up in the light. Here are my personal deductions from the harmony of this region: sky: blue, blue-green. Distance: blue-rose. Vegetation: the whole range of ochres, greens and blues. Rocks, earth, houses, tree trunks, etc.: the gamut of orange and ochre-red. No vermilion, since vivid green is excluded. This leads us straight to the palette of Cezanne! No fusion of colour, no intermediary tints, planes all of equal importance, of the same value. This countryside displays its colours like a beautiful rug, opposing each other in geometric designs which sometimes are monotonous. The landscape is not mastered without deep study and, perhaps for this very reason, it is to my mind the cradle of classic art.'

When we definitely leave the Mediterranean region the contrast is striking, as for example in the Cevennes when we reach the other slope of the mountains. Or again above Thueyts in Ardèche, when we have left the last chestnut trees behind we find ourselves on an undulating plateau. The long lines of the background to the west are barely accentuated, but to the north-east the daring summits of the Gerbier des Joncs and the Mézenc, real mountains, stand out against the horizon. We feel we are in another country, another climate. The colours are different, lighter, very soft and attenuated. The sky is pale, and dotted over with clouds that have come from the west to remind us of the distant presence of the ocean. The Mediterranean, though actually nearer, is forgotten. The south wind still sends its heavy rains, but this is as far as it goes. The west also sends a fine rain every day which wets everything but does no harm. Everything is deliciously green, and yellow is the basic colour. The villages are crude in tone with red roofs, giving an impression of a central plateau and mountains.

THE MEDITERRANEAN

The cultivation consists of a poor crop of cereals, but there are good pastures over which rove a fine herd of animals tethered with yokes. Here we have one of the great boundary lines of France.

Our painters have made a wonderful inventory of these colours, and as our writers have done almost as well both in their way have made an invaluable contribution to Mediterranean geography. An interesting piece of research that still remains to be done would be to gather from our literature any notes of Mediterranean colourings set down by the authors. In this way our authors would put at the service of our geographers their careful observations and innumerable discoveries.

There is another province which is still in great measure unexplored, that of odours and smells. I do not believe that smells have ever been dealt with geographically, not at any rate beyond a few stray unconnected notes which have not been handled on any really scientific scale. Yet it is an established fact that certain regions, certain countries and certain civilisations have a definite smell of their own. Kipling, in a pamphlet entitled *Voyages and Smells*, declared that he could detect two essential characteristics in the smell of a country: the type of wood that the people burnt, and the condiments they used in the kitchen. To this must be added the odour of clothes, and in nature the smell of the sap in grasses and trees.

Certain distinctions and interesting shades of meaning arise from these observations. For us the coast of Provence is associated with the scents exhaled by the shrubs and grasses of the *garrigue*, from the odour of fig trees basking in the sun, the smoke of smouldering leaves of olivewood, and the distant presence of forest fires. All that is typically Mediterranean, but there are other things too. People who have often visited the eastern Mediterranean associate that region with the odour of the fat of lambs' tails, and with meat cooked over aloewood, this being of course the domain of the kitchen. At Constantinople the smell of the eastern Mediterranean conflicts with odours that have drifted down from Russia. Towards Cairo and also towards Jerusalem a subtle perfume from the desert pervades the caravans and the nomads' robes.

These make up the odour of the East or of Africa, mingled with the exquisite perfumes and the smell of decay which come from the depths of Asia. This odour is wafted back towards the western Mediterranean, reaches our ports in Provence, and penetrates a little inland before coming to an end. I wonder where these far-reaching whiffs link up with other subtle smells? At Avignon perhaps, but hardly further north. Certainly not at Valence, nor at Tournon.

A geographic sketch of Mediterranean odours would be very difficult to make, but it certainly would be interesting. As in the case of colours, it would mean turning to literature where they are by no means overlooked. Our boundary lines certainly would not be as clear as customs barriers. There would be hesitations, haloes, temporary extensions like tides, perhaps even far-off enclosures and intricacies. But one would obtain the outline of something of real geographic significance.

III

Once again we have come to the conclusion that the Mediterranean countryside is a unit. It is everywhere the same, disclosing the same characteristics, and dragging after it the unity of its civilisation. In these surroundings, as indeed on any coast, it is the sea itself which is the centre of attraction. On every side this sea is surrounded by countries which look towards it and conjure up its name. It is their centre of gravity, even when their political capitals like Madrid, Paris or Cairo, lie outside the region. The individuality of the Mediterranean is even reinforced by its contrast to the regions beyond its boundary. By its very opposition to the oceans, to the deserts, and to the geological structure of the continental masses, the Mediterranean expresses itself in its own way.

No other country is more educative. To the learned, it teaches a sense of proportion. To the painter, a sense of line and composition. To the writer, construction. 'It would be impossible,' says M. Paul Morand, 'to write a badly constructed book alongside this deep blue sea.' Then M. Paul Valéry in his turn: 'Studying the shadows projected by the sun must have led to the first observation in geometry, that which we call projective. No one would have thought of it under a sky eternally veiled, any more than we in the north would ever have worked out a method of measuring time.'[1] This lesson is derived from the classics, in spite of the attraction that the romantics have always exerted over this marvellous southern sea. It is there that the classics were born. When they have been threatened and contested everywhere else, perhaps it is there that they will find their last refuge and fortress.

[1] PAUL VALÉRY, *Variété III*, p. 260.

CHAPTER VI

THE MEDITERRANEAN RACE

WE have arrived at the point where we must consider man himself. Let us now study the human element, and take into account the various races which have entered into the composition of the people of the Mediterranean region.

Groups of human beings are above all geographical groups — tribes, peoples and nations. They can be approached from several points of view, notably physical characteristics, language and civilisation. The anthropologist concentrates on race, the linguist on language, and the ethnographer on customs. Although each keeps fairly well to his own sphere, there is nevertheless constant confusion between race, language and culture. People often use the term 'race' when they mean language or civilisation.

A race, according to Littré, is 'a group of individuals belonging to the same species, having a common origin and similar characteristics which are transmitted from one generation to another' — obviously a purely anthropological group. A people on the other hand, is 'a multitude of men belonging to the same country, and living under the same laws'. According to these definitions to refer to the French race would not make sense, but we can say the French people. Nor must we talk of the Celtic, Slav or Aryan races, for the Celts, the Slavs and the Germans are properly speaking peoples. In any case there have been so many mixtures, and so much interbreeding, that a pure race no longer exists. In fact mingling of the human race has been universal and prolonged throughout the centuries. Some people even contend that there are no races at all. Yet one does observe that certain racial types occur more frequently in this or that group. This permits us to determine racial areas where the home is a more convincing indication than any

THE MEDITERRANEAN

frontier could be. Consequently it is important to study the races, for the human factor seen from the anthropological angle remains an essential element of civilisation.

11

Basing our study on the more recent findings of the anthropologists, let us now try to discover the far distant origins of the people of the Mediterranean zone, which go back throughout the centuries.

From the later palaeolithic age, it is possible to distinguish two dolichocephalic races in Europe, the more southerly one being the ancestor of the Mediterranean race of the present time. Then in the mesolithic or stone age a new race, brachycephalic this time, insinuated itself into a corner between the two others. After that in the north and north-east we trace the invasion of a final race which again was dolichocephalic. In this way at an infinitely remote period, the basic tripartite division occurred, consisting of the Mediterraneans in the south, the Alpines in the centre and the Nordics in the north. Later intermingling has not blotted out these distinctions, which are still so essential that they are found in every classification (Fig. 9, p. 93).

It is interesting here to link up Europe with Asia. According to the most recent studies and hypotheses, the three European races were the result of three great parallel ethnical currents moving out of central Asia. They concentrated and mingled to some extent on our small continent, which became more and more crowded as time went on. Thus the white race is a Euro-Asiatic entity. It is divided into three horizontal bands more or less parallel with each other: the Nordic race, dolichocephalic and fair-skinned; the Alpine race, brachycephalic; and the Mediterranean race, dolichocephalic and olive-skinned though incontestably white like the other two.

A great brown race, according to the classifications of Ripley and of Haddon, spread from Indo-China as far as the Mediterranean, passing through India, Arabia and Egypt. These people

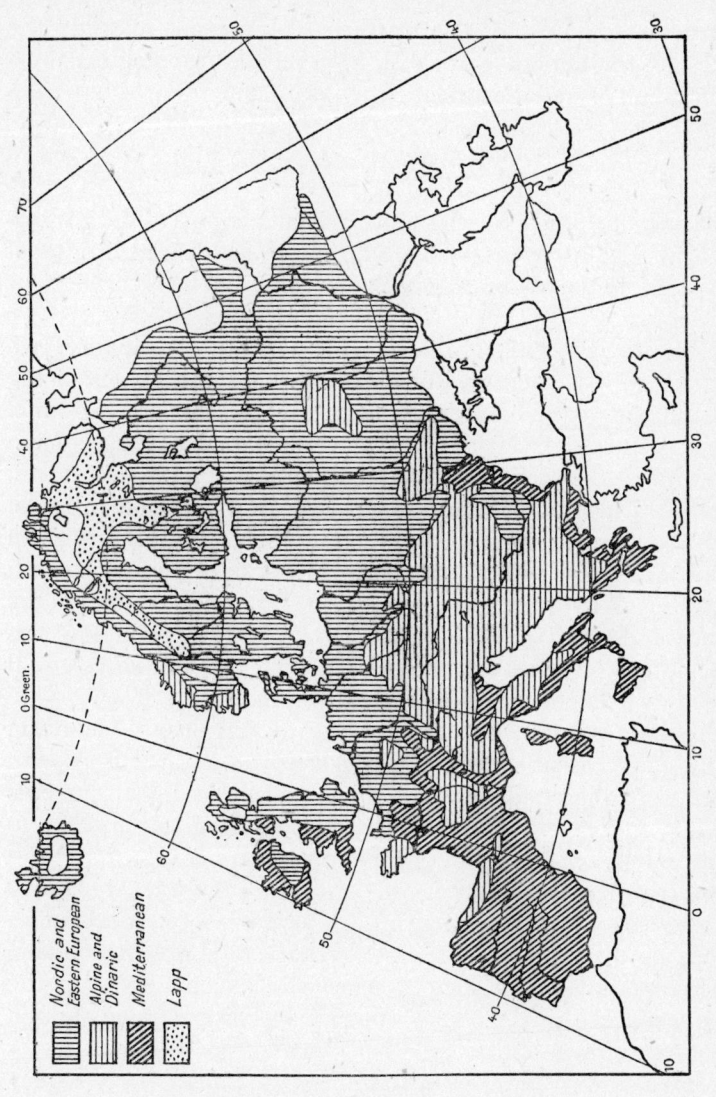

FIG. 9. THE GEOGRAPHICAL DISTRIBUTION OF THE EUROPEAN RACES ACCORDING TO VON EICKSTEDT

THE MEDITERRANEAN

may be more bronzed in colour than the others, but they are nevertheless an authentic section of the white race. With their wavy hair they are quite distinct from the yellow race who have smooth hair and yellow skin, and the black race with their

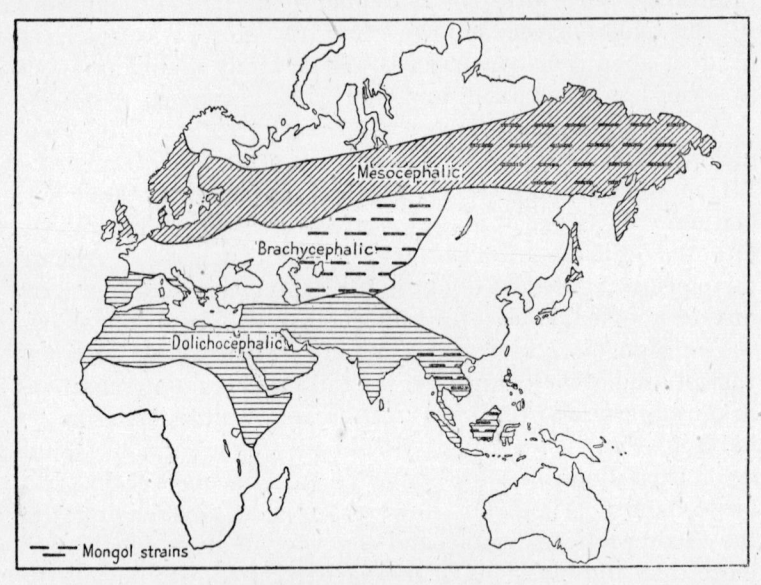

FIG. 10. THE DISTRIBUTION OF THE RACES OF THE OLD WORLD ACCORDING TO HADDON

curly hair. The Mediterraneans form the most westerly section of this racial family, whose axis runs from east to west. They are closely inbred with their European neighbours, the Alpines and the Nordics. On the other hand Persia and the Himalayas form a barrier separating the Asiatic section of this brown race from the Chinese (Fig. 10).

The Mediterranean race seems to trace its foundation to the peoples whom we find at the beginning of recorded history: Egyptians, Libyans, Phoenicians, Pelasgians, Aegeans, Iberians, Ligurians and Etruscans. After that people belonging to the Alpine or to the Nordic races descended from the north towards

THE MEDITERRANEAN RACE

the Mediterranean. Some of these invaders were more Nordic and others more Alpine, but as they were rather mixed it is difficult to determine their race exactly. In any case they brought in a new element, and then melted into the ancient substratum which already was complex to the point of confusion. All this makes us very modest about our knowledge, and very prudent about our conclusions. We certainly should be wiser to speak less about races and more about peoples.

We can distinguish, however, several great historical movements. The Nordics split up into the Celts, the Gauls, the Belgae, the Germans, the Goths, the Franks and the Normans, and more particularly in the eastern or central Mediterranean into the Acheans, the Dorians and the Umbrians. Among these people figured no doubt many Alpine elements who were further mingled, at least towards the east, with the Mongolians.

The Mongolian race sent a mighty tide of people as far as Europe, and of these the Turks are the authentic representatives in the eastern Mediterranean. As for the Arabian invasion, let me emphasise the fact that it did not add any new racial factor; it consisted simply of white Semitic people, members of the great brown family. The Arabs, however, spread a new religion over the Mediterranean region, and a new civilisation which even more than the invasions of the barbarians contributed to the decline of the Roman Empire.

As a result of these successive waves, new stock was continually added to the original basic elements, and it is this which led to the foundation of the Mediterranean race. Let us now try to detect its outstanding characteristics.

III

The Mediterraneans possess their own physical characteristics which distinguish them from the black, yellow or other sections of the white race.

They are not tall — 5 feet $2\frac{1}{2}$-$3\frac{1}{2}$ inches on an average — and the Alpine brachycephalic people are even shorter. They are

THE MEDITERRANEAN

delicately built, loosely knit, light, with an easy graceful frame. The dolichocephalic skull is the general rule, tending towards the mesocephalic, with index generally less than 75. The face is elongated vertically and also from the front towards the back. The nose is straight and prominent, rather larger than that of the Nordics, and one often comes across the so-called Grecian profile in which the nose and forehead are in line.

The skin is a swarthy white, but even when dark it has nothing in common with the black of the negro, being simply tinged with brown as a result of exposure to the sun. The hair is wavy and usually dark, while the eyes are brown or black. On the whole the type is elegant and harmonious, and although it has no very marked characteristics it is sufficiently definite to assign it a domain of its own geographically.

This domain in a general way is the Mediterranean zone itself in Africa, in western Asia and in Europe. The exceptions are the Dinaric Adriatic region where the race is Alpine, and the Anatolian Peninsula where the race is Mongolian. Let us say then — though this list is indicative rather than final — that the domain extends over northern Africa (Berbers, Arabs and Egyptians), the Iberian Peninsula, the French and Italian coasts of the Mediterranean, Italy south of Rome, part of Greece, and the islands and shores of the eastern Mediterranean. To this must be added sub-sections on the Atlantic coast of France between the rivers Gironde and Loire, on the southern coasts of England and Ireland, and in various parts of continental Europe.

It would seem that at a certain period the Mediterraneans spread themselves fairly far towards the north, to be thrown back again towards the south by later invasions coming from the east. The Mediterraneans who inhabit the Atlantic coast also are dolichocephalic, rather taller and of heavier build, with higher cheekbones, nose straight and prominent, skin pale, and hair black, straight and shiny. This Mediterranean element has formed a barrier against Alpine and Nordic influence in the section of the continent lying to the extreme west.

THE MEDITERRANEAN RACE

In France — as one can easily see in the excellent charts in the Atlas de France (Map No. 79) — the Mediterraneans of Provence or Languedoc may be distinguished by their figure from the people of the central plain or the Alps (Drôme and High Alps) who definitely are smaller. They are also distinguished by their dolichocephalic heads from the strongly brachycephalic groups of the regions of Lozère, Upper Loire, Aveyron, Cantal, Tarn and Garonne, and Lot and Garonne. In a general way the skin, hair and eyes are dark in the south of France and fair in the north. According to this one may speak of a French Mediterranean race as comprising a section of the Mediterranean race proper (eastern Pyrenees, Aude, Hérault, part of Gard, estuary of the Rhône, Vaucluse, Var, Maritime Alps and a section of the Atlantic). In spite of the many invasions from the north, this has race maintained its own individuality throughout the centuries.

No picture is complete without its frame. In order to understand the Mediterranean race we must also consider the races inhabiting the neighbouring regions.

Immediately to the north of it we find the Alpine race. These people are small in stature (less than 5 feet 4 inches), brachycephalic (index over 85), with high forehead, face round and wide, hair and eyes brown tending to black, skin of an intermediary colour between that of the Mediterraneans and the Nordics. The Dinaric race, a branch of the Alpine race, has similar characteristics, but is rather taller, extremely brachycephalic, with the nose often prominent or even hooked and the hair fine and black. The Alpine race occupies a strip of country lying east and west across Europe between the Mediterraneans and the Nordics; the Dinaric branch is to be found chiefly in the Balkans, Austria, Switzerland and northern Italy. The individuality of this Alpine race is often confused with that of the Celts, who are simply a people. The Alpines and the Celts do coincide, however, in certain cases, in fact it would not be too much to say that just as the Celts have a psychology of their own, the Alpine race has a definite psychology also. In the same way the psychology of the man of the mountains is one thing, but the psychology of the man of the coast is quite another.

The characteristics of the Nordic race are well established. These people are tall and well proportioned in stature, head dolichocephalic with features harmoniously developed, cheekbones not specially prominent, face long and narrow, nose prominent, profile regular, hair fine and usually fair and light, skin fair and pink, reddening rather than tanning in the sun. These characteristics have received considerable publicity in the United States and also in Europe, which always lays emphasis on the contrast between the peoples of southern and northern Europe.

In this case the racial domains are not contiguous. The Nordic comprises all the north and north-west of the continent extending south to the point where these people meet and mingle with the Alpines. The purest types are to be found in Scandinavia, in north-west Germany, on the east coast of England and Scotland, in Normandy and in many parts of northern and eastern France. Unlike the Mediterraneans the Nordics seem to have arrived in Europe rather late, actually towards the end of the neolithic period of the iron age. The many invasions bearing southward carried them to the shores of the Mediterranean, where their presence is still easily discerned, for their characteristics are so well defined. Their influence must have been important, for there is a distinct Nordic individuality, made up of certain qualities centring on their genius for action and organisation.

There is no direct contact between the Mediterranean and the yellow race. Yet to be exact, contact does exist with a certain race, Mongol or Touranian, which is derived from the yellow race and should be classified at least indirectly with it. The Mongols and the Touranians are short in build (less than 5 feet 4 inches), strongly brachycephalic (index over 85), with the drooping eyelids we associate with the Mongols, high cheekbones and nose rather flat. The Touranian is distinguished by being rather taller, and having a more prominent nose. In general

THE MEDITERRANEAN RACE

the Mongol or Touranian region comprises central Asia, Turkistan, Persia, the Crimean Peninsula, the districts inhabited by the Turcomans and the Tartars, and the Turks of Anatolia and of Europe. It is through the Turcomans that the Mediterraneans come into contact with the yellow race.

We must also take into account the exchange of population which took place in the eastern Mediterranean shortly after the close of the first Great War, when the Turks went back to western Asia from which the Greeks had been evacuated. We must also remember that a clearly defined racial border-line exists between the Turks and the Arabs; the latter are white Semites while the former belong to the great Mongol family. In this case the race marks a fundamental distinction which is further accentuated by the difference between their civilisations.

Finally, towards the south we find a fringe of negro population but it is a long way off on the other side of the desert of Sahara. It is not sufficient to describe a negro simply by saying he is black, any more than it is enough to say that a giant is big. The essential characteristic of his type may not be his colour. Some coloured people are very black, such as the Ethiopians who are not negroes. The black race is tall, with wide shoulders and a narrow torso. The dolichocephalic tendency is accentuated, with marked prognathism. The lips are thick and protruding, the nose is flat and the hair woolly and crinkled. Their domain is Africa south of the desert, but not Mediterranean Africa. To this we must add the branches of the black race inhabiting the Sudan and the Upper Nile, the Dravidians of India and the Melanesians of the Pacific (Fig. 11, p. 100).

We have thus traced right around the Mediterranean race a framework of other races. To what extent are these neighbouring people influential?

IV

In the Mediterranean region external racial influences have often made themselves felt. These influences have penetrated

THE MEDITERRANEAN

in the form of invasions of whole peoples, or by slow peaceful migrations or infiltrations, but also by means of an infinite variety of travellers, pedlars and salesmen, tourists and winter

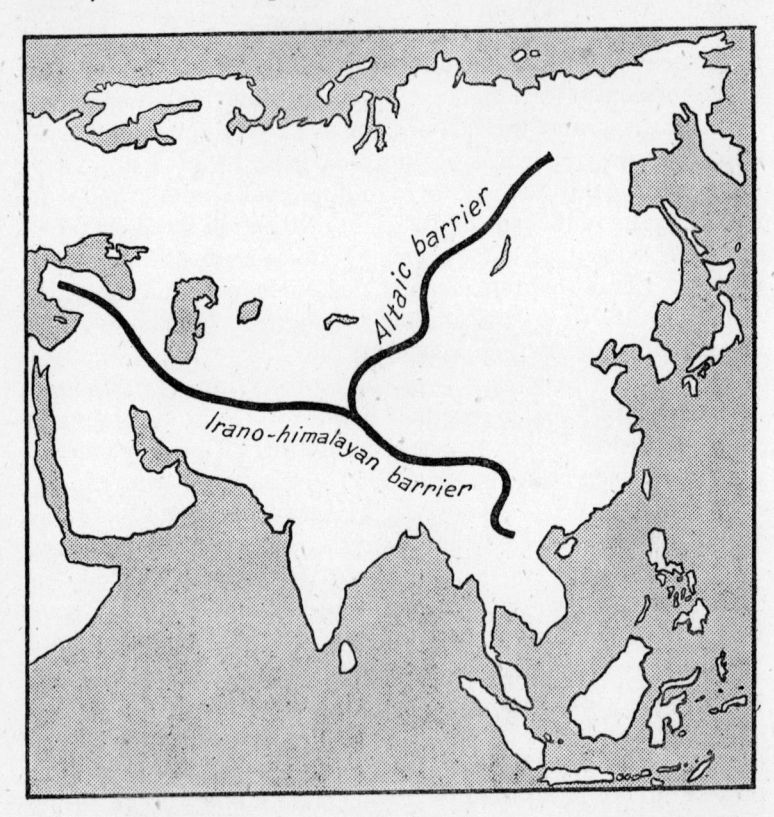

FIG. 11. THE THREE ZONES FROM WHICH PEOPLE OF THE PRESENT DAY ORIGINATED, ACCORDING TO THE FRENCH ENCYCLOPAEDIA

visitors, to which we must today add summer visitors as well. Some of the invasions have not amounted to very much, in fact although historically renowned they have meant the displacement of not more than a few tens of thousands of people. Nor have they always been racially homogeneous, nor even as a

THE MEDITERRANEAN RACE

rule. It is sometimes difficult to say whether a certain invasion was Nordic or Alpine. Or again, how are we to classify the Gauls?

Nevertheless these movements of population invariably did bring new physical elements into the Mediterranean zone as well as new customs, languages and ideas along with the importation of arms and utensils — all new in relation to the past. Often the newcomer was transformed as he adapted himself to the conditions of his new surroundings and climate, and submitted to the influence of an ancient civilisation upon which he imposed himself or into which he insinuated himself as an intruder. The Mediterranean is thus filled with non-Mediterraneans who, though assimilated, leave the imprint of their individuality on the region.

The various penetrations of the Nordics into the Mediterranean region have all left their mark. Since the beginning of recorded history, without even considering earlier movements, there have been numerous invasions of this origin — into Spain, France, Italy, Greece and Asia Minor. For example among the Greeks ancient elements from the north are superimposed on elements of a still earlier Mediterranean type. Yet it is difficult if not impossible to assign to each one the place from which his ancestors came originally. The German, the Gaul and the Slav are difficult to distinguish, and even if we do know that certain customs and forms of civilisation — less refined it is true — have come from the north, we still cannot assign to them the race to which they belonged. The Nordics, however, possess one essential quality in which they take special pride, and that is their sense of leadership, as they say in the United States. Thus the addition of this blood from the north has undoubtedly infused into the Mediterranean people a firmness of character which probably they could not have acquired otherwise.

Although there certainly was an Alpine penetration, it is difficult to locate it exactly. The Slav element, so widely represented in the Balkans and notably in Greece, is undoubtedly

THE MEDITERRANEAN

Alpine to a great extent. The same may be said of the human elements from the central plains of France who today hold such an important place in our Mediterranean Languedoc, or of the Italians who peopled Lombardy. Whatever their racial origin may have been, these imported elements brought with them psychologically a conception of earnest patient work, a spirit of thrift, and persistence in individual effort. The Mediterraneans, who have long since adapted themselves to their surroundings, are more brilliant but less solid than the newcomers, who in this respect have brought with them a valuable asset.

Should we speak of a Mongolian penetration? The Turks, although they are not Mediterranean in either race or culture, have always lived on the margin of the Mediterranean. They are not a seafaring people; they belong to the continent, to the land. If they are not soldiers they are farmers or cattle raisers, but they are not traders. Commerce is not their strong point, and they have never taken more than passing interest in the economic development of the eastern Mediterranean although they have dominated it for over five centuries. However, thanks to their useful and solid qualities they have made a fine contribution. They are honest and serious, with an ingenuous simplicity which at the time is rather shrewd. They are upright in character, which distinguishes them from certain neighbouring races who are inclined to push their wit and ingenuity too far.

Have the negroes penetrated at all? It would be an exaggeration to place any importance on their influence, but it would be wrong to say that it did not exist. On both sides of northern Africa — in Egypt and Morocco — a certain amount of infiltration is perceptible, but on the whole the Mediterranean is incontestably a domain of the white race.

As for the Arabs, their influence, let me repeat, is the influence not of a race but of a civilisation, and their effect on the Mediterranean has been enormous. Up to the seventh century the Mediterranean, according to Jules Sion, was the *mare nostrum* of the Romans, their own or almost their own familiar sea, around which they grouped people of their own culture and tradition. Since the beginning of Islam, however, the Mediterranean has acquired a dual personality. It has received from the East a brilliant civilisation, and has submitted to this influence as distinct from the Western civilisation to which it essentially belongs.

V

As our analysis shows, the Mediterranean, as a result of its privileged geographical position, has developed a human element of such refinement that it has contributed more than any other region in the world to the spread of civilisation. It has witnessed the birth of some of the greatest efforts of the human race: the civilisations of Egypt, Greece, Rome and Arabia, not to mention the European contributions of medieval and modern times. One certainly could not write the history of civilisation without assigning an outstanding role to the Mediterranean countries.

This spread of civilisation coincided with a certain period of human evolution. It began with the neolithic age and continued until the present time. It has always been linked up with the reign of the tool or other means of individual work — or at any rate work carried on by relatively small groups. During the last century and above all since the last war, the machine has introduced into the world a new conception of production. Production today is no longer dependent on the sinews of men and animals, but is based on motive power, on the elementary energy of nature.

In the new age of the human race which has only just begun, it is not certain that the Mediterranean will be as well able to play the same role as in the past. This is now a burning question, and one which is causing the Mediterraneans acute anxiety. The natural conditions of their environment have not changed: their soil and climate are always the same. But their economic and technical problems do not lend themselves to the same solutions. Other regions and other continents are better placed for mass production than the Mediterranean is, with its individualism and attention to detail. The question is whether the inhabitants of this highly exceptional part of the planet are capable of finding, and of applying solutions which will permit them to adapt themselves to circumstances which were unknown to earlier civilisations.

What then is the racial value of the Mediterranean people? It is a section of the white race, as we have said, and it possesses the essential qualities of that race. It is also a section of the brown race which is only the most southerly offshoot in Europe, Asia and Africa of the great Euro-Asiatic race. It is not a question of a race which is purely brown or purely Mediterranean. First of all because there is no such thing, and secondly because favourable additions have later improved the initial foundation. All interbreeding is not good, but sometimes it is excellent. We can occasionally discern crosses between different races which certainly have led to civilisation and progress.

The result has generally been good when Nordics and Alpines descending towards the coast have brought their own characteristics of work, patience and sense of organisation into the more civilised region into which they have penetrated. Probably neither Greece nor even Rome itself could have developed as they did without this foreign strain, and in both cases the infusion of new blood proved to be excellent. The allurement of the southern sky has always had a powerful effect on the people of central and northern Europe, who are enchanted by so much light and charm in comparison with the harsh climate of their own countries. Even after they have been transplanted into the milder climate, their original virtues continue for some time, and this has undoubtedly benefited and hastened the flowering of one of the most magnificent civilisations in the world.

But mingling and intrusion have not always had this happy effect. On occasion the wave of immigration has taken the brutal form of a tide-gate, destroying everything as it passes, and leaving nothing but disorder and ruin in its wake. In ancient times certain countries never recovered from such catastrophes. Also, the mingling of foreign strains may not be beneficial if the incoming race is too different. Now the qualities of the Turks are indisputable, and the West has always been ready to recognise them — and yet they are not adapted to the Mediterranean region. As a result the Turks and Mediterraneans even after several centuries have gone their separate ways, like two currents of different nature which even long proximity could not combine. The Mediterraneans returned to the sea, the Touranians went back towards the high tablelands, and in one way and another they became friends only after they had separated. From this we conclude that the Mediterranean, although eminently fitted for all the members of the white race, is not and cannot be a suitable region for the development either of the yellow race or of the races derived from it.

Finally, we must add that a foreign strain if it introduces a new racial factor does not exert its effect indefinitely nor even for long. As a rule the immigrant is not assimilated during his own lifetime. Until his death he belongs to his native land, and his qualities and shortcomings are scarcely modified by his new life. His son still talks of his father's country, but it is simply a memory. The grandson finally belongs definitely to the environment into which he is born and where his father was born before him. If any influence remains of the country of origin it is almost intangible, for by this time it is dominated by the new surroundings. In three generations the cycle is completed. This is the law of immigrant assimilation, and one may observe its unchanging rhythm in America. Today the Mediterranean is bristling with problems which are remarkably like those that the United States had to deal with when Europe was pouring forth her people overseas.

In the course of this short study we have been confronted with two ideas: race and civilisation. These two ideas are associated and yet distinct, and we must not confuse them.

THE MEDITERRANEAN

No doubt the various races have served as vehicles for the civilisations, but the latter in their turn have transformed the races. As a result we have the Mediterranean world with its vigorous individuality. But we cannot understand its psychology until we have studied the circumstances and the conditions in which the problems of production have been tackled and overcome.

CHAPTER VII

TRADITIONAL AGRICULTURAL METHODS

THE unity which the Mediterranean region owes to its climate is reflected in its agriculture. Agriculture here is the result of long tradition. It is adapted to the environment, perpetuated by history and fixed by routine. But alongside methods which date back to the time of Homer, types of culture are being developed today in accordance with modern technical and commercial requirements. One might be in California! Here we have two worlds in very truth, or shall we say two stages? — that of the tool, still almost neolithic, which we find in the small holdings where people live their own lives with little contact with the outside world; and at the same time specialised industrial agriculture based on the machine and in contact with world markets. The second stage represents a recent expansion and is very important. Nevertheless it is the first stage which even today is typical of the Mediterranean.

I

M. Paul Valéry writes: 'The work of man opposes its will to build, opposes the voluntary almost rebellious effort which is typical of our race, against the tendency which geological Nature has expressed in the collapse of its crashing landslips.'[1] Now M. Paul Morand: 'The Mediterranean stands in opposition to the desert. Its entire history is the struggle of settled people against the nomads, of the stone house against the hut, of deep-rooted vines against the hasty harvests of bands of horsemen, of the Latins against the Slavs, of farmers against

[1] PAUL VALÉRY, *Variété III*, p. 247.

cattle rangers, of stone against sand.'[1] Such are then the circumstances underlying this economy: the effect of the intelligent human will on nature, the struggle against the desert, and the establishment of a settled civilisation.

Natural conditions here are at once favourable and difficult. Alongside a limited amount of excellent land are vast stretches of mediocre soil, which is cultivated if necessary but not if the good soil will suffice. Apart from a few exceptional cases this rules out large-scale cultivation. Prairie lands are rare, and many of them have not been used as they were unhealthy, as some of them indeed still are. It is the hillsides which offer the best possibilities, even in spite of the 'curse of the limestone soils', and the ever-recurring landslides which means that to maintain the soil is quite as difficult as to reclaim it. As a result the Mediterranean farmer's work is never finished. He must turn his land into artificial terraces, and then guard them against storms to prevent his fertilised soil from being carried away by floods of water.

Nature here makes no free gifts and allows no relaxation, but on the other hand she reduces human needs to a minimum. In this happy country where comfort is not very important, the famous Anglo-Saxon standard of living is not the measure of either civilisation or of happiness. But the Mediterranean should never reckon up his work. The continual handicap of set-backs and interruptions, of long treks on foot across the mountains to reach some isolated and distant piece of land, would exasperate a Taylor or a Ford. This is not the atmosphere of mass production, which is efficient only when it is monotonous and continuous.

A second enemy is drought. If it is necessary to drain off the excessive moisture after a deluge of rain, it is equally necessary to preserve this precious water in order to apply it exactly where it is needed. In this climate the effects of irrigation are miraculous, in fact with irrigation one can accomplish almost anything. The mountains, like a great reservoir, are always

[1] Paul Morand, *Méditerranée, Mer des Surprises*, p. 13.

AGRICULTURAL METHODS

at hand. Let me quote a couple of lines from La Fontaine, lines which every Mediterranean knows by heart:

There, under the rugged rocks, near a crystal spring,
Is a spot respected by the winds and avoided by the sun. ...

I have in mind a spring of water which has been taken captive and led down the face of the slope by means of a little canal. In this way the beneficent water is brought to the gardens of the entire region and is preserved in the many reservoirs that reflect the sky.

Owing to the geographic structure, irrigation can be carried out only in a limited way and on a small scale. Those who make use of it form associations on a basis of equality, and from these associations little democratic communities are born. In the Mediterranean a whole civilisation is founded on irrigation, which has been organised by local initiative and ingenuity. These associations are in complete contrast to the impressive public works of ancient Egypt — the planned economy of the Nile — and also to undertakings in the United States and our own North Africa. What we need is not a bucolic La Fontaine, but an epic poet to sing the praises of the formidable modern dams.

When there is not enough irrigation to withstand the effect of drought, the Mediterranean resorts to a type of cultivation that requires less moisture, such as olive trees and vines whose deep roots seek out water far below the surface. These trees are planted well apart, so that each one has its own share of moisture, but the soil must be kept malleable at the cost of endless work. This kind of cultivation has always been practised in these parts and actually it is the same as what is called 'dry farming' in certain of the newer non-European countries.

The contrast between the sea and the mountains close by has a direct effect on this exploitation. The possibilities of this contrast had already occurred to people in ancient times, to men like Thucydides and Strabon. The plains, although too hot in the summer, are quite habitable in the winter, while life in the mountains is exactly the reverse. So the population moves 'from the blue bed to the brown', to quote the Vicar of Wakefield. Thus the low-lying country, which is fertile but often unhealthy, can be used in two ways, and the same applies to the higher levels which are temporarily adapted to

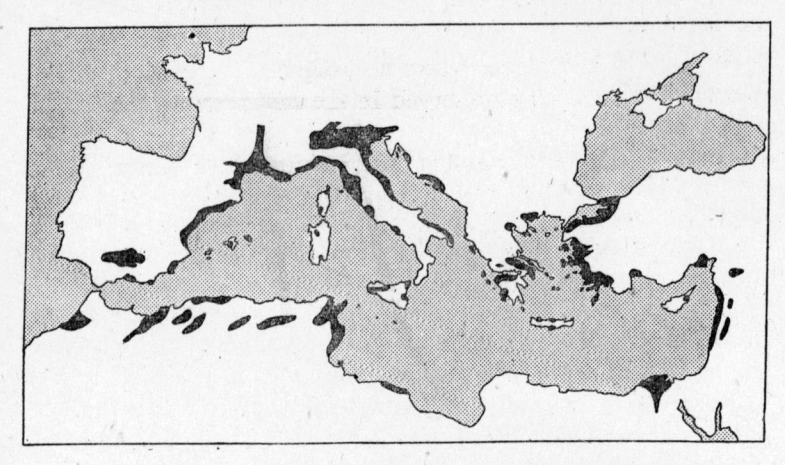

FIG. 12. REGIONS OF INTENSIVE CULTIVATION

sheep-raising. People settle on the slopes which are usually intensively cultivated. Thus the lands under agriculture tend to be dispersed and scattered, with limited patches of excellent soil and great stretches of *garrigues*, *maquis* and mountains (Fig. 12). M. Parain, in his book, *Méditerranée*, refers in this sense to the Island of Murter in Dalmatia. This island consists of a belt of pasture-land, another of terraces covered with olive trees and grape-vines, and a third of loess soil where wheat and barley are grown. The three zones, however, are not adjoining.

In view of this natural structure the fields are generally irregular, and sometimes slope steeply. It is difficult if not impossible to use agricultural machinery, so farmers turn to the old traditional implements, though less by force of habit than by necessity. The equipment used in the Mediterranean is very simple, almost archaic. One feels that under these skies time does not pass as quickly as it does elsewhere. It is in other lands that machines have been invented to revolutionise agriculture.

Mediterranean equipment, however, is perfectly suited to the local needs: the cultivator (*araire*), which is a plough without wheels for the light soils; the hoe; the pick and the pickaxe; and the sickle which is more useful than the scythe. The harrow is hardly used at all. The grain is often trodden out by oxen on threshing floors of hard earth.

One can well appreciate the emotion and respect that these old implements arouse in a writer like M. Paul Morand: '... One of these ploughs for poor land, and like all agricultural implements in the Midi it is minute. Compared with tools in northern countries these look like toys, except that they are worn out, grooved, and chipped by flints, thus betraying the toil to which they have been put....'[1]

These tools, primitive and simple as they are, are the best for this kind of soil. Farm hands are to be had, and they are essential to this work, for when horses and oxen are lacking the machine cannot take their place. The American has no patience with this sort of thing, because he is spoiled. He despises the back bender, the man who bends over the earth and makes a great effort with the swing of his spade — but what else can he do? The woodsman in the fable preferred his wooden hatchet to the golden hatchet which Mercury offered to him. And the god had to agree with him, for this was the wisdom of very old countries.

This kind of cultivation is well suited to a certain geographical environment and to a certain period of human work. It is moreover self-supporting. In other words it produces what the family needs — food, clothing and tools, and in consequence it introduces variety into production. It also requires ingenuity, attention to detail, foresight, close observation and

[1] PAUL MORAND, *L'Homme Pressé*, p. 33.

THE MEDITERRANEAN

ease in adaptation. For this a craftsman rather than a peasant is needed.

All this is very different from conditions in northern countries. Here the farmer spends little on his land but he asks little in return. He spends more time on his garden — on his vines and his fruit trees — than he does on his cattle and his fields. It is not so much hard heavy work that is needed as unremitting daily effort. What would the Mediterranean small holder not do for his little property! He is not lazy, and he knows exactly what he will get in return for his work. The American objects to any interference with his standard of living, and the farmer of northern France also is inclined to look down on so much frugality, even in comparison with his stern necessities of comfort. Yet, with his wine, his oil, his fruit trees and his radiant sky, the Mediterranean is happy enough. He is free in spirit, too free perhaps. Let us not forget Archimedes, who regretted that he had to make use of his knowledge in his research for practical solutions.

The danger of a civilisation of this kind is that it presupposes an accumulation of sufficient capital to build terraces, and to lay out irrigation canals and drainage. These things take time to establish, they are difficult to maintain and must be constantly protected against the elements. Then, let us say, a prolonged period of war, political disorders or anarchy takes place. The terraces crumble, the canals lie empty, stagnant water brings malaria, and finally the land threatens to return to the desert. Many regions — Crete for example after the Doric invasion — have never been able to recover from such disasters. This is a lesson which we should not forget.

II

The zones of production, according to the authoritative statement of Vidal de la Blache, fall into four categories: first, we have the mountain slopes overlooking the sea; the small plains with intensive irrigation; the larger plains, and finally the mountains. The coastal slope is the most typically Mediterranean, so let us generalise on this aspect.

With regards to this coastal slope Vidal de la Blache writes as follows: 'The mountain slopes along certain parts of the Mediterranean coast are formed like an espalier or fruit-wall, so that man need only trim the gradients to his needs. In addition there are also little sandy beaches sheltered from the mistral and the north winds. These little beaches are within reach of one another, so communication is easy thanks to the mild winds and the temperate climate. The beaches, which are virtually shut in by the mountains that rise up steeply from the shore, are suitable for fishing and coastal trading. The way that the fruit groves are always in contact with the sea is a truly Mediterranean characteristic.'[1]

Three stages are superimposed on this slope — we are here at the very heart of Mediterranean life. First, we have the coast proper up to about 600 feet above sea level, with little harbours leading a double life as town and marine. Then we reach the stage of bushes and shrubs, between 600 and 1800 feet up, and a series of little fortified villages at the 900-foot level. Here we find olive trees, vineyards, groves of citrus fruit, vegetables and flowers. Finally at from 1800 to 2700 feet above the sea the mountains begin in earnest, with well-exposed slopes where chestnut trees flourish. This might be described as an extension of the Mediterranean.

According to Vidal de la Blache, it is not the field but the garden that is the pivot of the people's existence. The shrub benefits by deep roots that assure it a minimum of moisture, and irrigation can be concentrated on anything grown under the trees in the fruit groves. On the plantations — incidentally the term was used by Xenophon — the principal crops are olives, citrus fruits such as orange, lemon and grapefruit, grapes, figs, almonds and apricots, flowers, and early fruits and vegetables.

As for grains, in the past there was an important crop of

[1] VIDAL DE LA BLACHE, *Principes de Géographie Humaine*, pp. 81-95.

cereals here — wheat, maize and above all barley — but these are gradually being replaced by more specialised cultivation. Stock-raising is restricted to the smaller type of animals, such as goats and sheep which do not require much attention. There are few beef cattle except on the Mediterranean borders of central Europe, for the farmers are short of fodder and other by-products which would serve as cattle feed. They get around this difficulty, however, by sending their livestock to other pastures, but with the progress being made in intensive cultivation cattle-raising is steadily diminishing in importance. This is not the traditional atmosphere of the peasant, but something rather different. One might describe the Mediterranean as a craftsman of the land.

The small intensively irrigated plains give an exceptional yield, and when good soil happens to lie near an abundant spring of water the result is as miraculous as an oasis. A typical example is the Mediterranean coast of Spain with its *huertas*. The people there irrigate their gardens by the Arabian method and produce a wonderful crop of beans, lucerne and ground-nuts under the orange and apricot trees. Scores of small holders, all sharing the life-giving water, conform to the traditional rights that have held good for centuries. The *huerta* at Valence, which can be seen from the cathedral, is the finest and therefore is always cited as an example. It is 600 square miles in area, and supports 300,000 inhabitants. This population which is extraordinarily dense for a European countryside recalls the enormously rich deltas of Asia. In France the plains of the eastern Pyrenees may be classed in this category, and also those of the island of Hyères, and again those at Palermo in Sicily, but generally the plains are less fertile than this.

The drawback of the plains, and especially the great Mediterranean plains, is that they are naturally unhealthy, for water accumulates in the depressions and becomes stagnant and malarial. Consequently farmers have always hesitated to settle there, and when they do they turn up as briefly as

AGRICULTURAL METHODS

possible at seed-time and harvest. There are no small holdings on the lowlands, where the land is divided into large estates, and the people live in groups on the mountain slopes nearby. This has long been the tradition on the Campagna at Rome, and on the plains of the Basilicata, Thessaly and Andalusia.

On the other hand these same plains, once drained and made healthy, will give excellent yields, as has been proved in Languedoc and Mitidja, and also in Spain and Italy. As many low-lying districts, especially in the eastern Mediterranean, still have to be drained they are sparsely populated. The Mediterranean plain is thus less typical of the region than the mountain slopes are, and all the more so when the plain is suitable for cultivation, for then modern methods are usually adopted. We then have a single crop — cereals, olives or grapes — cultivated mechanically according to the principles of mass production. This does not belong to the Mediterranean tradition, but to a more recent stage which we shall discuss in greater detail in the next chapter.

Finally we come to the mountains which form a framework so close to the shore that they can usually be seen from the sea. Here we have tablelands, forests, pastures and bubbling springs. When we arrive from a sun-baked district everything seems deliciously rustic and green, and we can well understand how charmed the ancients must have been with these poetic meadows and hillsides. And yet these same mountains can be as stony and sterile as the desert.

As a matter of fact they should not be included in the Mediterranean region, but they do belong essentially to its economic life. For one thing trade is always going on between the upper and lower levels, and people go back and forth as they and their livestock migrate to fresh pastures each season. Then again the mountains act as a reservoir, for the springs provide the water needed for the irrigation carried out on the slopes. Thus the mountains are always there, distant perhaps but still very real, for they allow districts to be cultivated which otherwise would remain wild and unproductive.

THE MEDITERRANEAN

In some splendid passages Vidal de la Blache describes the effect exerted on the Mediterranean by the lofty summits which on fine days can be discerned on the horizon — mountain peaks such as the Canigou, the Mulhacen and Olympia which act as patrons for the districts of Perpignan, Grenade and Brousse. The very fact that they are nearby is important geographically, and socially hardly less so. The mountain streams go tumbling down the slopes, and with them also go the men who bring their vigorous energy and forgotten highland customs down to the heart of the plains, where life is almost too easy. In the reverse sense vegetation ascends the valleys as far up as the mountain passes. Languages and ideas, doctrines and theories also go as well — in fact all the things which bring the civilisation of the coast up-country to a point where resistance is as strong as penetration. At that point an equilibrium is established, and the influence of the low countries comes to an end. Thus it is impossible to understand the Mediterranean without its neighbouring and complementary framework of mountains.

III

Small holdings producing a variety of things for the local market, and consequent independence of world economy — such are the essential characteristics of this way of life which dates back to earliest times. Xenophon's analysis still applies, and in the gardens of our own Riviera one meditates on Virgil. The sobriety of the Greeks of the classical epoch who lived on bread, vegetables and fruits, recall the simplicity of the Mediterranean cultivators of the present day. The barbarians of the north who gorge on butter and dairy produce seem coarse to these men of the south, who have reduced their wants to a strict minimum, and thus have more time to spend on other things.

The Arabs certainly have introduced new types of cultivation such as sarrasin, asparagus, hemp, flax, mulberries, saffron, rice, palms, lemons, oranges, coffee, cotton and sugar-cane.

AGRICULTURAL METHODS

They have also made the very most of irrigation, and have turned the whole region into a vast orchard. Nevertheless the spirit has remained the same, and even under Arabian influence the Mediterranean has not changed its nature. Cultivation has continued on much the same lines across the centuries. Even today life in the countryside of Greece or Lebanon recalls the agriculture of the days of Homer. There are few regions in the world which have remained throughout the centuries so completely true to themselves. This does not exclude progress, but even progress continues in the same groove for fundamentally the conditions of production have not changed.

Meanwhile the modern world is discarding and revolutionising its methods of production. How will this industrial revolution affect the Mediterranean?

CHAPTER VIII

INDUSTRIALISED CULTIVATION

In the Mediterranean region it has long been the tradition for small holders to grow a variety of crops, principally on the mountain slopes and hillsides, and to dispose of them in the local market. However, a new stage, one that definitely belongs to our time, is introducing entirely different conditions. Cultivation is now specialised, and carried out on the plains instead of on the hillsides. With the change has come large-scale mass production based on the machine, and serving not the local market but great international trade outlets.

Various complex reasons have led an important proportion of the Mediterranean growers to adopt modern methods. For one thing technical progress and new means of transport now bring the produce on to the market after a very brief delay, with the result that grapes, flowers, early vegetables and citrus fruits have acquired a completely new value. On the other hand, the increasing competition of extra-European countries, which grow wheat and other grains on an enormous scale, discourages production on poor soil which obviously is unsuitable. Under these circumstances the Mediterraneans, like everyone else, have been forced to specialise and turn to mass production. It is true that suitable soil is the exception in the Mediterranean region, and therefore the arguments against the new methods are plausible enough. Nevertheless, with this reservation, mechanised and serialised cultivation is winning out, and not only on the large well-drained plains but even on some of the mountain slopes. The movement is all the more accentuated in the newer districts such as French North Africa, where the spirit of the settler is naturally eager for progress and ready to take risks. As a result the whole Mediterranean set-up is being rapidly transformed, becoming

INDUSTRIALISED CULTIVATION

so American at times that one has the impression of another California.

Let us take for example the modern vineyards on the plains of Languedoc. In the eighteenth century a variety of crops were still grown in this region — wheat under the olive trees, mulberries on the mountain slopes, grapes for quality wines on the stony hillsides, beef cattle on the pastures in the lowlands along the coast, and pines for timber on the sandy shore. The first step was taken when the *garrigues* were reclaimed and turned into vineyards by the small landowners at the time of the French Revolution. Then, after the Restoration and especially after the Second Empire, wine production expanded enormously and thanks to the railway found the world markets it required. The change was sensational. At the end of the Monarchy Languedoc was still a countryside like another, producing a variety of crops without thought of specialisation. The State even seemed to dread the change towards the single crop, for in 1730 a decree had been passed forbidding further extension of the vineyards. Nevertheless, in a few decades the plains were already becoming an ocean of vines. By 1824 there were 250,000 acres planted with grapes, and in 1860 no less than 550,000. The phylloxera epidemic between 1869 and 1878 proved to be only a temporary set-back, though it certainly was disastrous enough at the time. However, the wine producers of the south survived, rejuvenated and in possession of all the latest ideas and technical methods.

Today the plain between the Cevennes and the Mediterranean looks like an enormous sea of grape-vines, with other types of cultivation standing out like a few rare islands. The market towns are as big as cities, and the great wine companies use the most up-to-date equipment. Specialisation is now complete. Grain, fodder and all kinds of vegetables are imported into the district as they are no longer grown there. Even the people concentrate their entire attention on wine production and the industries connected with it. The towns, too, are devoted to the wine trade — such places as Béziers,

THE MEDITERRANEAN

Narbonne, Lézignan, and also to a great extent Nîmes, Montpellier and Perpignan.

As both the social structure and the psychology of the people changed at the same time, this might almost be a new country with a different centre of gravity. In the eighteenth century the small holders began to reclaim and cultivate the hillsides, but in the nineteenth century they came down on to the plains, more especially as the plague of phylloxera spared the lowlands which were subject to floods. The result was an important colonisation of the low country, but this time under the auspices of the great landowners and wealthy companies. These concerns were backed with enough capital to be able to afford cement vats, wine-cellars and narrow-gauge railways, in fact all the equipment necessary for their huge undertakings. In the beginning these undertakings were capitalist in character, but the Wine-Growers' Co-operative which has developed more recently has now diverted the profit from this investment to the small owner himself.

Doubtless there are serious drawbacks to this wholesale transformation of an entire country to mass production, for the wine is no longer of fine quality. An enormous market is essential, and if it is lacking a sharp depression occurs, for the wine-growers have no other type of cultivation to tide them over. Salaried staffs have to be found, people who are craftsmen rather than peasants in character. Finally, the single crop industry has had the same effect here as it has everywhere else: speculation as in the United States, and the temptation when things go badly to turn to a planned economy. Apart from this, the work accomplished has been magnificent. At any rate it has proved that the Mediterraneans are as capable as anyone, perhaps even better, at adapting themselves to the conditions of modern large-scale production.

I have taken Languedoc as a typical example, but I could just as well have described the Algerian wine industry. Up to the time of the phylloxera epidemic our little colonies in North Africa had kept to their traditional French customs.

INDUSTRIALISED CULTIVATION

They raised cereals and cattle, and were independent and self-supporting. Their wheat gave them a living but did not make them rich. Then, at the end of the nineteenth century, France looked across the Mediterranean to Algeria, and decided to develop a new wine industry on a large scale. Both land and labour were cheaper in Africa than in the mother country, and plenty of space was available. From about 100,000 acres under vines in 1879 the area rose to about a million after 1930; today wines and their by-products make up 40 per cent of the Algerian exports.

We should not accept out of hand the dilemma between quality wine and that turned out by mass production. Although quantity no doubt is a decisive factor, there is still a third eventuality to be taken into account, that of good quality which briefly is not just the same as quality. A division of labour is taking place. The best wines are made on the slopes, and according to tradition this is a family affair. On the plains, however, it is a case of mass production, and the two methods are being pursued simultaneously. Meanwhile a great reforming wind has blown up. These people are in touch with the outside world, not only with their own home market but with international outlets as well. They are no longer thinking individually and in a small way, and that in itself is a revolution.

If it is a question of early vegetables, of fruits or flowers, the evolution — revolution if you prefer the term — is exactly the same. In the past sales were narrowly limited geographically. Transport was slow, for goods were carried in carts or even on men's backs. Perishable commodities had a restricted market, though actually limited less by circumstances of production than of distribution. Rapid transport by rail today brings these same perishable goods within reach of distant markets. Commodities that yesterday could not be used in any great quantity can henceforth be exported to urban markets, which have so changed in character that they have become insatiable for foodstuffs and luxuries.

Thus, cultivated on lines of mass production and with

specialisation carried to extremes, fruits, flowers and early vegetables have become one of the great sources of wealth in the Mediterranean, and not only in Europe but also in Africa and even in Asia. Wherever irrigation exists, the possibilities are endless. So the grain field is receding into the background, to give place to market gardens and fruit groves. This change has led to semi-industrial development with skilled management, packing and freezing plants, and suitable railway stations. Our own Midi, Italy, Spain, Palestine, North Africa and the admirable Mitidja have become new Californias on this side of the world. This revolution is the work of the white race. The Arabs were in the region first, but they follow rather than originate, so credit must go the Western civilisation.

Even the olive tree has been dragged into this revolution. On the slopes which only yesterday were covered with plantations, people have been cutting down the trees. No doubt they would go on doing so, in France at any rate, had a decree not been passed prohibiting such a massacre. In the past the olive tree played an important role in Mediterranean economy. When the plains were unhealthy it could always be grown on the slopes, and the oil was used in all sorts of ways. Nowadays the olive trees are neglected or badly kept up. This is partly owing to lack of labour, because other types of cultivation are more remunerative, or finally, because it is impossible to compete with other kinds of oil.

So the olive-oil industry is declining, but that is not the whole story. On the other side of the Mediterranean this same olive tree is being mass produced over vast territories, making use of the latest technical equipment. With labour cheap and plentiful, it is already regarded as one of the main types of future cultivation. In southern Tunisia in the neighbourhood of Sfax the olive is still on probation, but this also is a district into which the twentieth century has introduced its own methods.

INDUSTRIALISED CULTIVATION

II

Thus an entire section of the life of the Mediterranean is being industrialised. Wine-making is now an industry, first owing to its equipment and then by its methods of distribution. The manufacture of wines is becoming more and more of an industrial operation, whether it is carried on in the great caves of the co-operatives, or by the big companies which buy the grapes and then convert them. In these circumstances it is the industrialist who is the real producer. The wine tends to become the raw material of large commercial undertakings with a national outlet, which work and distribute on a colossal scale. Even their transportation is becoming more and more specialised, for they now have their own type of motor trucks, their own tank cars on the trains, and their own tank boats which are not unlike oil tankers.

The case of early fruits and vegetables is not unlike wine production, for here again the small holder exists alongside the large company. Nevertheless, the family undertakings are still the most efficient, with their hard work, patience and ingenuity, though these people too are resorting to the same methods as in California. They are installing much the same modern equipment — scientific irrigation, special pipe-lines, and stoves in the orchards and citrus groves to guard against frost. In the same way the Mediterranean growers have resorted to collective organisation in the distribution of their produce, as they must have rapid transportation by rail or boat. They have also introduced standardised qualities in order to serve their markets effectively. All this obviously demands an effort beyond the scope of the individual.

In the flower industry to which these modern conditions also apply, the product tends in many cases to become simply the raw material for the manufacture of perfumes, another trade that has been highly industrialised. As a result we have great horticultural enterprises, heavily capitalised and with large staffs of salaried workers directing their intensive cultivation

THE MEDITERRANEAN

on scientific lines. Alongside them are medium-sized undertakings, family businesses concentrating on flowers and employing very little outside help. Many of these small growers cultivate fruit and vegetables as well as flowers, and still keep a few olive trees as in the past.

In one way or another the earlier individualism, with the man working on his own account, is threatened. He is in danger of being supplanted by the large organisation which can meet the imperious demands of the new industrial epoch. 'Anarchic individual', is the way the American technocrats refer to the Mediterraneans, as they sing the praises of the mass man of the future. But the Mediterranean individualist can defend himself, and moreover, defend himself by means of the collective organisation which he has already accepted, and which he is already developing to the best of his ability. In other cases the producer is losing his independence. A sort of proletariat is now being constituted, semi-rural, semi-manufacturer, and very different in tradition from the small holders in both psychology and instinctive reactions.

The change extends even into public life, into the relations of the individual and the State. Formerly irrigation was merely a local matter, cheap enough, and requiring little investment on the part of the users. The cultivator with little equipment could always manage. If the harvest was bad he simply waited, for he did not have to worry about either banks or distant markets. Today, especially in the regions that have been recently developed, irrigation is a tremendous affair with great dams and long complicated canals. It is now a matter of public works which demand scientific technique and important capital investment, and this means recourse to powerful capitalistic enterprises or to collective undertakings. As in the United States the cultivator is involved in much greater initial costs than was the case in the past. His equipment is more complex and more expensive, and the cost of his irrigation is heavier. If he borrows from the bank his risk increases when the market declines, and then the burden of his debt makes itself felt.

INDUSTRIALISED CULTIVATION

The social repercussions of this revolution are widespread. The economic horizon has suddenly been extended, and henceforth it will be impossible to keep to the particularism of the small establishment traditionally closed in on itself. The grower must now consider placing his products not only on the national market, but on the most distant international markets as well. A depression in our day, whether it be at Shanghai or Buenos Aires, sooner or later affects the producers of wine, fruits or flowers in the Old World. The Mediterranean grower, whether he likes it or not, cannot ignore what is happening abroad, even on the other side of the globe. His forebears, who were never harassed with such anxieties, were happier in this respect.

In these conditions, the cultivator of today has to be much better educated technically and economically than his father was. He has to be a bit of a mechanic, and many of the things that he must have at his finger tips can only be learned at school or from specialised trade journals. In the same way he must be better informed generally, not only about his own country but also about other countries, in order to know where he can sell his produce and in what currency. And that is not all. In a time when administration penetrates everywhere, he must know all about administrative methods and practice, so that even when he is away out on the *maquis* he can still hold his own with the proceedings of the syndicate, the co-operative, and even with planned economy. Only yesterday the Mediterranean civilisation was a human affair, based on the influence that his trade exerted upon the worker, and on a man's individual reasoning in the light of his own personal experience. The coming civilisation is based upon information dealing collectively and extensively with a subject, information such as is to be obtained in the trade journals, on the radio, and by contact with the trade associations. The new kind of agriculture is wider in scope, but it brings with it less harmony and peace of mind, and in the long run less wisdom.

The effect on social relations is already apparent. The

personal and familiar character of social contacts were simpler in the past, more dignified, and more on the same level in that they were independent of differences in fortune. Daudet stressed this point in several of his novels about the Midi, and the same was true of Spain and Italy. People went about with ease, and mixed with other classes of society without losing their own natural simplicity of manner.

The new economic regime on the contrary increases these differences. Today the manager or director of a great establishment is very far removed from his salaried employees who are almost anonymous, and who have even given up hope of ever becoming their own masters. The old Greek tradition of social struggle is threatening to return in our day, and although it may be rejuvenated in form it is equally serious. It is also striking to find that the brake exerted by local tradition is becoming less influential. In the present period of rapid communications in which dialects, clothes, houses, furniture, personal tools and even customs are indistinguishable between one canton, one department or even one province and the next, mingling and intermarriage intensifies this sort of standardisation. The North African labourer knows Europe and the Italian emigrant knows America. Without denying that our birthplace exerts a deep influence on every one of us, isolation could not exist as completely as in the past, even if former methods of cultivation continued to prevail.

III

The Mediterranean by its structure and its atmosphere does not lend itself naturally to the industrial revolution, and yet it is becoming adapted to it. It is true that this applies chiefly to the regions where the geographic personality of the Mediterranean world is less marked, that is to say on the vast open plains, in the wide valleys penetrating into the interior, and in districts where the structure of the country belongs more to

INDUSTRIALISED CULTIVATION

the continent than to the coast. The present revolution in methods is in the nature of things, and yet the greater part of the Mediterranean world remains attached to the past and its ways. Alongside genuine twentieth-century achievements, archaism still persists. The charm of this sea is that it is so old and so young at the same time.

CHAPTER IX

PROPERTY AND OWNERSHIP

It is only when we analyse its system of property and ownership that we really penetrate the economic and social character of the Mediterranean. This region is one of small proprietors, but the large estate also claims an important place. Doubtless this is partly a legacy from the past, but it is also the logical consequence of the new methods of production. In studying the balance between the various types of property, one realises that it is the reflection of natural causes, and also that the present development is as it should be.

A small property is established and maintained when the soil is fertile, but above all when the harvests are regular and plentiful enough to allow a family to live off the land. Ten or twelve acres is a good average, but when the soil is very rich, as it is in the irrigated gardens and fruit groves, even a couple of acres will suffice. Also the small holder can set himself up if tools and hard work seem to be more vital than capital and machinery, or shall I say when initiative, patience and steady effort will succeed where the machine and mass production are of no use. Victory then goes to the man who can adapt himself and his work as required. The circumstances where the small holder cannot survive are equally clearly defined, however; he has a poor chance when the harvests are uncertain, notably in regions with little moisture and irregular rainfall. The cultivator is then at the mercy of a bad year which gets him into debt and may even land him with a mortgage. This soon leads to reorganisation on a much larger scale.

Within certain limits the conditions of the small holder are logically defined. The proprietor can work with his family, and eventually with the help of a day labourer or a friend; but his business must not be too big, nor yet too small. On the margin

PROPERTY AND OWNERSHIP

are various intermediary grades, such as when the property is in process of either formation or disintegration. A very small holder while he is getting himself established will continue to go out to work by the day, in other words he is both owner and wage-earner at the same time. In Languedoc it is the recognised custom for his employer to lend him tools, and eventually a horse. In the same way a cultivator who has got into debt will hire himself out by the day to try to get his affairs straightened out. There is a striking difference in psychology between the small independent proprietor, the labourer who is already a proprietor or who hopes to become one, and the farm hand who will never be a proprietor and knows it. In the French Midi distinct political climates indicate where each of these three types of worker is in the majority.

The geographical and topographical domain of the small property-owner in a general way covers, first, the irrigated zone along the coast; second, the small subdivided plains with good exposure and water supply; and third, the slopes, terraced hillsides and foothills of the mountains overlooking the sea. In the main the rivieras have been claimed by the small holder, and one needs only to wander around them to realise the extent to which the land has been divided up, and that here individualism and specialisation are the order of the day.

As for the larger pieces of property consisting of over fifty acres of good soil, and especially over 250 they become commercial undertakings when capital and machinery are the factors that count. For this purpose it is necessary to have broad surfaces which will lend themselves really well to agricultural machinery. Then on suitable land rich harvests can be obtained by mass production, as in the case of wheat, citrus fruits, and certain grape-vines that have a bountiful yield. For reasons which in a way are negative the large proprietor also takes over land where the small holder cannot succeed. For example, the unhealthy marshlands which were part of the big estates of ancient times are beyond the means of the little man, and so are the vast stretches of intensive cultivation which require solid financial backing owing to the irregularity of the harvests.

In the past the great Mediterranean estates were sometimes feudal, sometimes military, and sometimes the result of a political grant or concession. When they were too big they often led to an intermediate form of indirect exploitation, such as cultivating the poorest land on shares, leasing the better farms, or employing managers of the type of 'general farmers'. It has nearly always been necessary to resort to the large landowner — either a company or a wealthy financier — to develop land that has been neglected or has never been cultivated. Then again outside capital is needed to win back or assist in recovering property that has got out of hand during a crisis. This was the case, for example, after the phylloxera epidemic in the vineyards of Languedoc. Scientific and technical treatment on semi-industrial lines is needed on these occasions, and a large corps of work people must be employed.

This kind of exploitation is more suitable on the great plains than on the hillsides. In the French wine districts of Languedoc at the foot of the slopes, on the hillsides of Spain, Italy, Syria and especially in North Africa, the small holder simply carries on as usual. In the Mitidja the holdings were medium-sized in the beginning, but owing to their amazing success they have now become large, very large in fact.

All this has brought about a perfectly natural equilibrium, which exactly reflects the conditions of success and set-back. As both have their good points it is impossible to pronounce definitely in favour of either the small man or the great landed proprietor.

The little fellow has less technique, and makes less progress, but only he can develop the stubborn land as he regards it as his own personal affair. He does not count his own time as part of his costs, and he is willing to accept a very low standard of living in order to succeed. He works his land with a passionate devotion which is inspired by mysticism as much as by economy. In a sense man then becomes the creator of the land.

PROPERTY AND OWNERSHIP

With regard to the large estates, we must distinguish between the methods of the ancients and those of the present time. In those days the great proprietors held the mediocre land, and their methods of cultivation were inferior. The absentee landlord never was an instrument of progress, for he turned over the whole undertaking to managers who took little interest. The classic judgment, *latifundia perdidere Italiam*, became a by-word.

The large modern property is an entirely different affair. It is splendidly equipped, the last word in technique, and run by competent men who are keen on their work. Now they most certainly are instruments of progress. Languedoc, as I have said, reminds me of America, and the Mitidja recalls California even more vividly — actually it is superior in many ways to the set-up on the Pacific Coast.

There is a marked difference in the social structure of the communities of small holders and the large estates. One might even be watching two stages of evolution, one before and the other after the introduction of machinery. Once when I was in the Department of Le Gard I was asked to write a monograph on two 'communes' one of which, Ledignan, was situated at the foot of the Cevennes, or just beyond the plain; while the other, Saint Laurent d'Aygouze, was in the lowlands not far from Aigues-Mortes.

In its 254 voters the first community included 130 landowners or 51 per cent, 67 labourers or 26 per cent, and 26 craftsmen or 14 per cent. Out of 540 voters the second was divided into 264 landowners or 48 per cent, 264 labourers or 48 per cent, 9 craftsmen or 1.6 per cent. Thus on the one hand there was more variety on the slope, while on the other hand on the plain, where the social classes were sharply divided, the number of craftsmen, curiously enough, obviously had declined. Probably as production here was more highly mechanised and on a larger scale, the people looked to the neighbouring town and not locally for their equipment.

M. Parain gives figures for the vineyards of Bouffarik in

Algeria which also are significant: There a large estate of over 250 acres accounts for 58 per cent of the area; medium-sized properties of between 150 and 250 acres, 23 per cent; and various holdings of less than 150 acres, 19 per cent. In each case the owner seems to have made a satisfactory choice.

Thus from the technical point of view both the small and the large holdings have their advantages, but from the social point of view the small holding is better. The small holder populates the country and is attached to the land; he works hard, and best of all he is satisfied with his work. The peasant may even become so completely identified with his land that he loses all sense of political patriotism: he only knows the patriotism of the land. This is the case in Egypt, and also in other places where the peasant is not in the least interested in wars and conquests. If they will just leave him his field, he will be satisfied. Like the horse in the fable, he asks if he is expected to carry a double load. It remains to be seen whether the large estates can escape the agrarian struggles which overwhelmed them in the past. In spite of its beneficial role in modern development, the large estate is today meeting a new edition of these earlier difficulties in the form of strikes and class warfare between employers and employees.

11

The conditions under which the Mediterranean population has been built up are much the same as those which have determined the development of the district. Naturally people will settle wherever they find a means of existence, either directly by producing food for their own needs, or indirectly by exporting raw materials to serve as a medium of exchange. People are attracted to places where the geographical structure or the climate are suitable for the human race — that is to say plains, valleys or favourable slopes, hospitable shores, and regions with natural resources, such as coal for example. Similarly they flee spontaneously from districts, however rich,

PROPERTY AND OWNERSHIP

which are unhealthy, difficult of access, or where they do not feel secure.

What is to be had in the Mediterranean? There are many wealthy and prosperous regions here, but apart from a few exceptions they are never far from one of the deserts. These arid zones are seldom very extensive, as a matter of fact they are usually only isolated patches. On the other hand there are many districts, favourable in other respects, which are quite unsuitable for habitation. They may be unhealthy, and this difficulty is only now being overcome; or exposed to attack, for only yesterday insecurity was still a matter of tradition. Consequently the Mediterranean has been settled in a succession of densely crowded zones alongside others that are desert or semi-desert. This is indeed the 'anti-desert', as Paul Morand calls it!

In addition there are many anomalies, such as mediocre or poor land which is populated and should not be, and good land which should be populated and is not. These anomalies correct themselves of their own accord. One finds regular movements of population such as migration between the coast and the surrounding country, and emigration to more distant parts. In the modern world with its easy means of communication the Mediterraneans can always find refuge elsewhere. This was not so in the nineteenth century, nor even in the twentieth century, and although transportation certainly has improved technically, such changes and adaptations are still hardly more feasible.

One soon sees what is obsolete and primitive in this picture. As both coal and minerals are lacking, there is no hope of developing industry on any great scale, so people must live by means of agriculture, commerce and services. Furthermore, insecurity and unhealthy conditions, although eliminated elsewhere, still stubbornly persist here. Nothing could emphasise more clearly the contrast between central and western Europe and the Europe of the Mediterranean.

Let us now look at the sparsely populated districts. People

by tradition seem to have avoided the plains, which have always been malarial and lacking in security. Even today it is striking to find that the great plains of Spain, Italy and Greece are still comparatively empty. For example the seaboard of Tuscany boasts only thirty or forty inhabitants per square mile, which is much less than the average population of the rest of the country. Nor have the mountains attracted any permanent settlers either.

Let us not be tempted into day-dreaming about the idyllic Greece of the classical poets. Rather let us face up to the snow and ice of the rigorous mountain winters, winters which in our Cevennes continue a full six months, with snow which lasts and thaws only when spring is well advanced elsewhere. This explains the seasonal migration of the herds from the Spanish Sierras down to the plains of Andalusia, from the Alps to Provence, from the central highlands of France to Languedoc, and in Italy from the Abruzzi to the Roman Campagna, from the Apeninnes to the Basilicata, from the Pindus to Thessaly. The coast, at least the immediate shoreline, always seems to have been considered dangerous. In our Midi the invasions of the Moors have left many reminders of the way the people took refuge in the inaccessible mountain heights up-country, and established little fortified towns when they retreated from the sea.

On the other hand we find that people have concentrated on the coastal slopes of the mountains where, according to Vidal de la Blache, the life of the settlement is centred on the garden. The olive tree is typical of this country up to about 1200 feet above sea level, where the little walled towns are perched inaccessibly here and there on the peaks. Thucydides wrote about the different types of settlement: 'Recently when the sea was considered safer and people were better off, they formed these communities either on the coast or inland within the protection of the walled towns. They took possession of the isthmuses in order to carry on their commerce, and also to protect themselves from their neighbours as piracy continued on for many years. These towns, built well inland, are still there today.'[1]

In the communities along the various riviera coasts we find the town and marine collaborating in a way which is characteristic of the most densely populated parts of the Mediterranean, of places where there may be as many as four or five hundred inhabitants to the square mile. A little further inland is a strip of country which is even more thickly populated than the coast. This is the result of the co-operation of two distinct zones writes Vidal de la Blache. 'Here we have two different ways of life, seafaring and fruit-growing. This leads to a characteristic combination of Mediterranean life. Population and activity are concentrated on certain parts of the shore, while other parts are neglected as they are considered unsuitable.'[2]

The small irrigated plain, situated close to the mountain which acts as its reservoir, also attracts people in great numbers. There one finds settlements which are almost Asiatic in their density as they often have as many as 800 inhabitants per square mile. This is hardly European, not typical at any rate of a strictly agricultural district. We must not include the Delta of the Nile in this survey, for its population of some 14,000 inhabitants per square mile is less typical of the Mediterranean than of the great deltas of India or Indo-China. As for the large Mediterranean plains, once they have been drained and made healthy, nothing could be more perfectly adapted to cultivation on a large scale. As we have noted modern enterprise is already selecting them as its first choice.

Next let us consider the mountain slopes. In former times they were considered an admirable place to settle and make a living in preference to districts which, although they may have been more suitable, were avoided owing to their lack of security. When one explores the districts around 2500 to 3000 feet above sea level in our own Provence we are always coming across terraces that have fallen into ruins, and yet the old folk say they

---

[1] THUCYDIDES, *Peloponnesian Wars*, I, p. 7.
[2] VIDAL DE LA BLACHE, *Principes de Géographie Humaine*, p. 86.

can remember a time when crops were harvested there. We then go down from the heights which in normal times no one would ever have dreamed of cultivating. In the past people seem to have obeyed a law of gravitation which forced them to descend in much the same way as we have done.

For the same reasons which forced them to flee from the seacoast, they also congregated in the little towns, each of which once boasted an *acropolis* or an *oppidum*. For other but rather similar reasons the farmer never lived on a malarial countryside. He may have had to tramp a long way to get to his field, but in the evening he always went back to his home inside the town walls. The Mediterraneans are a gregarious people and they like to live huddled together in towns and villages. This dread of isolation is perhaps a distant survival of their earlier state of mind. Now that piracy has receded into the past and the plains have been made healthy, the farmer is at last tempted to settle down on the lowlands. Accordingly the towns are leaving the heights and being established down below. Today most the Alpine towns of the foothills still have a settlement up in the mountains where people once took refuge from danger, and a lower town in which the real life of the community is carried on.

For some reason the Mediterranean towns are always rebuilt on the same sites. The walled villages — little cities really — are full of vitality and vigour. In our day, however, it is the great ports which attract the excess population and become overcrowded. A Mediterranean town seems to be filled with a vivacious crowd, a plethora of human beings who give a wrong impression of the economic activity of the place.

Even from this brief analysis we have gathered that the Mediterranean has been populated in an entirely different way from continental Europe. In modern times it is the great unbroken plains of the continent and its industries based on coal that have attracted the people. A chart indicating the deposits of coal and the industrial centres coincides pretty well with a chart of the population. In the Mediterranean the only concentration that is at all comparable would be the great ports,

PROPERTY AND OWNERSHIP

which naturally become industrial centres as it is so easy to bring coal to them by boat. But here again we return to the same contrast of two different economic civilisations.

III

The fact that people come down from the heights and no longer wish to work is often interpreted as a sign of immorality. They wish to enjoy life, they say, and the energy of their ancestors does not appeal to them in the least!

This is less a decline than a struggle between two different conceptions of existence. In the past the ideal was frugality, linked up with work and individual effort. As no one knew anything else, people accepted without discussion a way of life which they thought could not be otherwise. Today they have discovered that it is quite possible to lead an easier existence, and it is no longer considered necessary to stay up in the mountains as it is no safer there than anywhere else. So they go down to the plains, and they emigrate.

Is this a lack of morality? Not at all, but rather substituting another kind of morality. This, however, has not yet come to pass, but still belongs to the future.

CHAPTER X

MINERAL RESOURCES

Two basic geological facts dominate the whole question of mineral and industrial development in the Mediterranean. In the first place this zone is contained entirely within the framework of the Alpine folds, and secondly the Hercynian chain lies outside the Mediterranean area. Now the Hercynian chain contains a wealth of coal and iron, hence its great possibilities for the development of metallurgy and industry generally. In these conditions the Mediterranean cannot expect any special industrial development, not at least while industry — heavy industry — continues to rely on coal. On the other hand considerable development of non-ferrous metals may eventually take place. This development may even become quite important, for the irruption in the Tertiary era produced resources which are by no means negligible, although of course not to be compared with the vast reserves of the Hercynian system.

In a world which is not dominated by iron and steel the Mediterranean can hold a notable place, even a leading place, and such was the case in ancient times. Our civilisation, however, is based on the overwhelming association of iron and coal, those two great sources of industrial, military and political power, and the Mediterranean is seriously handicapped in the international economic field as a result.

II

Let us first ascertain the place occupied by the Mediterranean in the world's production of various minerals and metals.

This table is significant (Fig. 13, p. 139) for although it emphasises the serious lack of coal, it also discloses appreciable reserves of non-ferrous metals, and important deposits of various

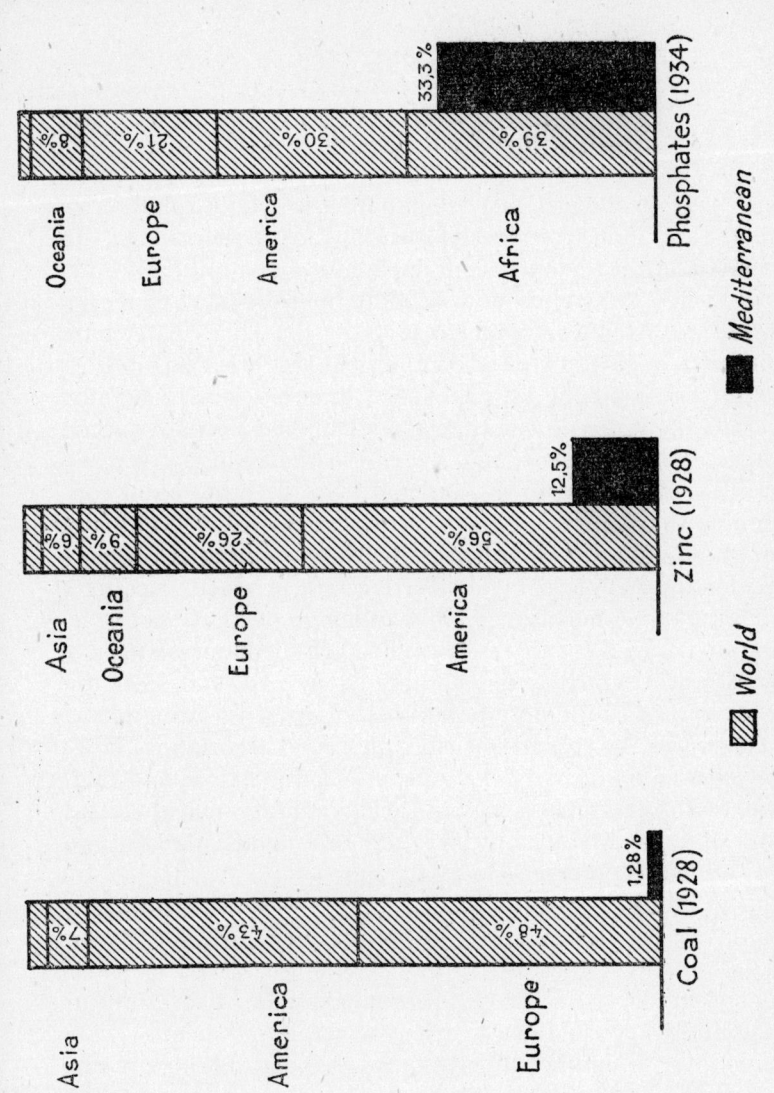

FIG. 13. THE MEDITERRANEAN'S SHARE IN THE WORLD PRODUCTION OF VARIOUS MINERALS

THE MEDITERRANEAN
PERCENTAGE OF WORLD PRODUCTION ON THE EVE OF THE WAR

		Mediterranean Production per cent
I	*Coal and Iron Ores*	
	Coal	1.28
	Iron	5.38
	Chromium	33.30
II	*Non-Ferrous Metals*	
	Lead ore	10.43
	Zinc ore	12.50
	Copper ore (metal content)	4.70
III	*Various Minerals*	
	Mercury	84.00–
	Sulphur	16.00
	Bauxite	30.00
	Phosphates	33.30
	Potassium	12.50
	Petroleum	.20

other minerals such as mercury, phosphates, bauxite and chromium. Finally, petrol, if not on the spot, is not far off.

Let us now run through the list in greater detail, if only to underline what is lacking. The most serious deficiency is coal, as production hardly attains 16 million tons a year out of the world production of 1250 millions. None of the Mediterranean countries produces any coal to speak of, and most of them have none at all (Fig. 14, p. 141). France mines about 50 million tons, but apart from the coal mined at Alès and the lignite at the mouth of the Rhône, it is chiefly in the north. The same applies to the 6 to 7 million tons in Spain and the 2 million tons in Turkey, which strictly speaking are not mined in the Mediterranean area at all. When we have noted about 2 million tons in Italy, 400,000 tons in Jugoslavia, 140,000 tons in Morocco, lignite at Cap Bon in Tunisia and also in Euboea in Greece, we may not have exhausted the subject but we have

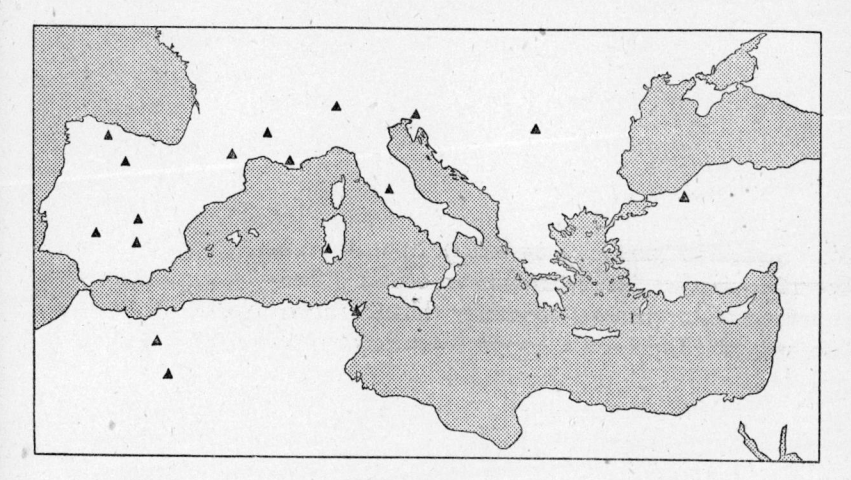

FIG. 14. COAL PRODUCTION IN THE MEDITERRANEAN

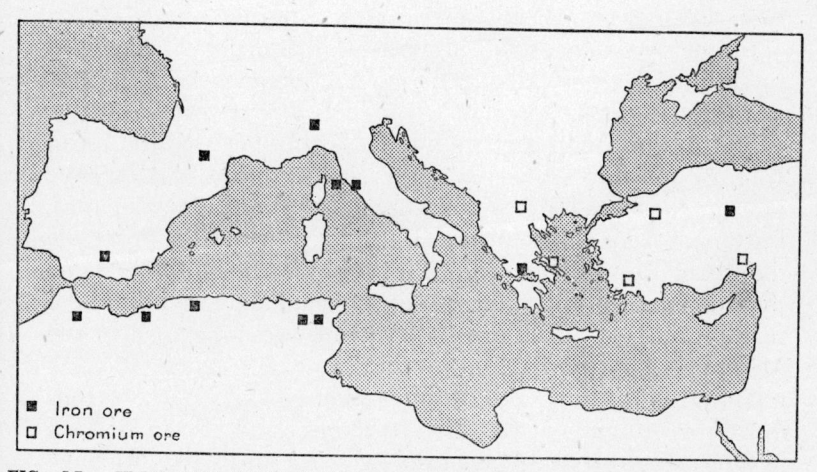

FIG. 15. IRON AND CHROMIUM ORE PRODUCTION IN THE MEDITERRANEAN

said all that is essential. If here and there interesting possibilities do exist, they do not amount to much.

When we come to iron ore the situation certainly is better, for production is about 7 million tons out of the world total of 130 millions. The principal producers are Algeria with 3 million tons (Ouenza, Mokta el Hadid and Tebessa), Tunisia with 800,000, Morocco with 1,150,000 (Spanish Rif), Italy with 950,000 (Island of Elba), Spain (Teruel, Cartagena, and Murcia on the Mediterranean coast). Many of these minerals are of such excellent quality that in future North Africa will likely stand high among the districts supplying raw materials to the metallurgical industries of Europe.

Conditions are even more favourable for chromium, of which the Mediterranean produces 300,000 tons out of the world supply of 900,000 tons. This comes chiefly from Turkey, Greece and Jugoslavia. Considering the importance of this mineral in the manufacture of special kinds of steel, one can hardly overestimate the value of these deposits (Fig. 15, p. 141).

In the case of non-ferrous metals, the Mediterranean again ranks well. Lead ore production, for example, amounts to 240,000 tons out of the world total of 2,300,000 tons (Fig. 16, p. 143):

		tons	
Jugoslavia	with	75,000	(Trepca)
Italy		60,000	(Sardinia, Lombardy and Tyrol in the Alps)
Spain		60,000	(Penarroya)
Morocco		30,000	(Touissit, Zellidja)
Tunisia		28,000	(Djebel Ressas, Sidi Amor)
Greece		20,000	(Laurium)
Turkey		10,000	(Balia)
France		6,000	(Alès)

The Mediterranean provides 375,000 tons of zinc ore out of 3 millions for the whole world (Fig. 16, p. 143):

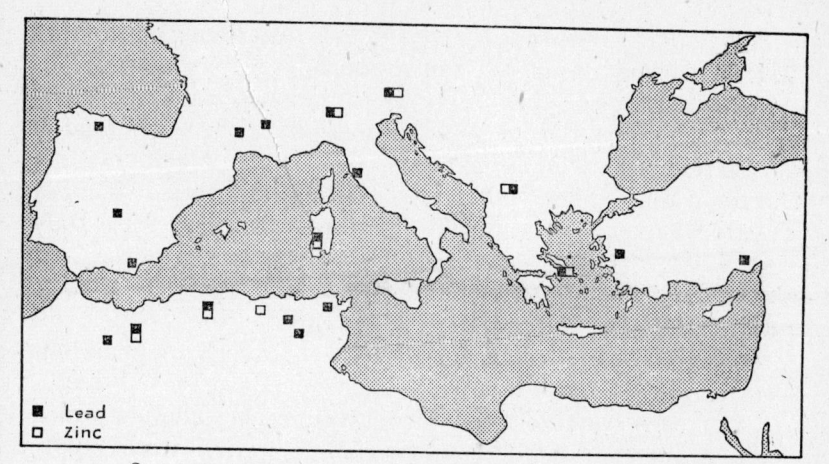

FIG. 16. LEAD AND ZINC PRODUCTION IN THE MEDITERRANEAN

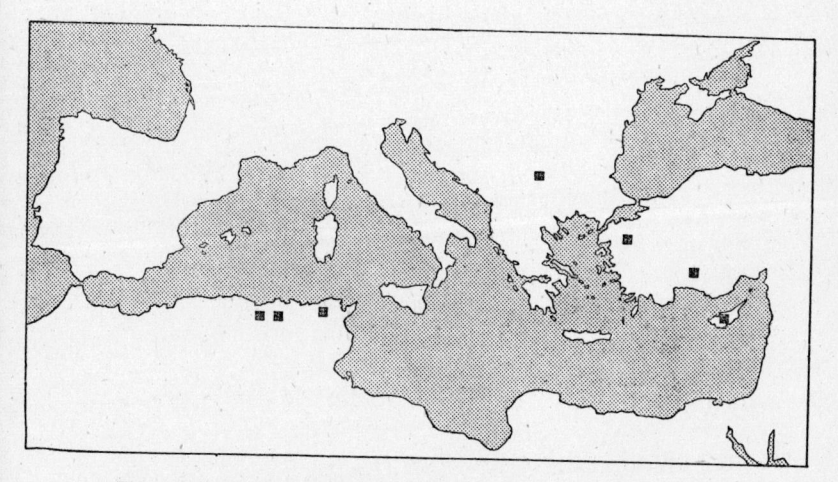

FIG. 17. COPPER PRODUCTION IN THE MEDITERRANEAN

THE MEDITERRANEAN

tons

Italy	with	180,000	(especially in Sardinia)
Jugoslavia		100,000	
Algeria		15,000	(Guergour)
Turkey		14,000	
Greece		10,000	(Laurium)
Morocco		3,000	(Haut-Guir)

As for copper ore (metal content), the figures are 80,000 tons out of 1,700,000 tons for the world (Fig. 17, p. 143):

tons

Jugoslavia	with	40,000	(Bor)
Cyprus		30,000	
Spain (Huelva pyrites)			
Turkey (Ergani, Murgul)			

Finally let us set down the impressive list of rich minerals of various kinds:

Mercury (cinnabar), 4200 tons out of a world total of 5000 tons, notably in Italy (2500 tons, Tuscan coast), Spain (1700 tons, Almaden, Grenada) (Fig. 18, p. 145).

Phosphates, 3 million tons out of a world total of 9 millions, notably in Tunisia (2 million tons, Gafsa), Morocco (1 million tons, Khouribga), Algeria (550,000 tons, neighbourhood of Tebessa) (Fig. 19, p. 145).

Potassium, 250,000 tons out of a world total of 2 millions, notably in Spain (250,000 tons, principally in Catalonia).

Sulphur, 350,000 tons out of a world total of 2,150,000 tons, almost entirely in Italy, but some also in Spain and Turkey (Fig. 20, p. 147).

Bauxite, 1,640,000 tons or about 35 per cent of the world total principally in France (700,000 tons, Les Baux, Var), Italy (400,000 tons), Jugoslavia (400,000 tons, Dalmatian coast), Greece (140,000 tons, Aegean Islands) (Fig. 21, p. 147).

We must devote a special paragraph to petrol (gasoline).

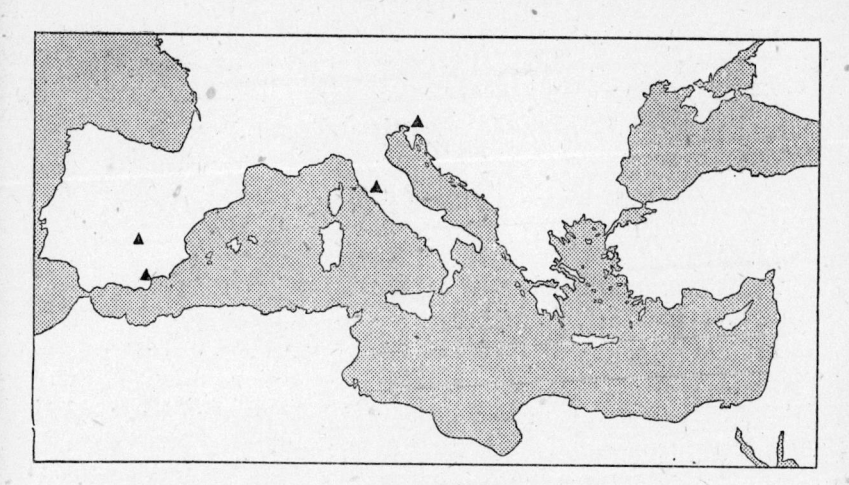

FIG. 18. MERCURY PRODUCTION IN THE MEDITERRANEAN

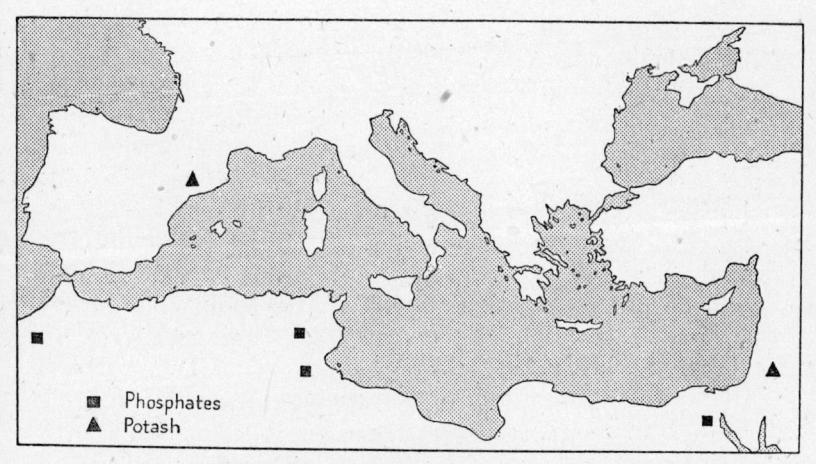

FIG. 19. PRODUCTION OF PHOSPHATES AND POTASH IN THE MEDITERRANEAN

Production in the Mediterranean is very limited indeed, being only 200,000 tons out of a world total of 200 millions, for although there is a little in Italy, Morocco and Egypt, it is not enough to matter. What is important, however, is that the pipe-line brings the Mesopotamian petrol to the coast of Syria and Palestine. It ends at Tripoli and Haifa, whence the precious liquid is easily distributed to the various Mediterranean ports (Fig. 22, p. 149).

This enumeration gives us a clear picture of the situation. It discloses elements of mineral wealth which if not extensive still are appreciable, and above all are varied. Nevertheless, the moment that we make comparisons the inferiority of the Mediterranean region is laid bare. Take coal, what is 16 million tons alongside an output of 400 millions in the United States, which in some years has even exceeded 600 millions? Or shall we say zinc (metal content)? Italy produces 90,000 tons, but the United States mines 420,000 tons and Australia 208,000. Or copper — Spain produces 24,000 tons and even this is not on the Mediterranean side of the peninsula, while the United States produces 515,000 and Chili 348,000! The great mining areas of the world lie elsewhere, and to aggravate the situation we are now in an era of mass production.

III

On the eve of the present war almost all the mineral production of the Mediterranean was exported in the form of raw materials. So far as any manufacturing processes were concerned, they seldom amounted to more than washing the ore or other equally simple operations. The reason was of course that there was not enough coal to carry the work of transformation any further.

Certain rare exceptions do exist, notably lead and zinc, but local conditions generally do not allow for the addition of foundries to the mines. No doubt in the case of minerals of light metal content like zinc, it is tempting to carry out at least

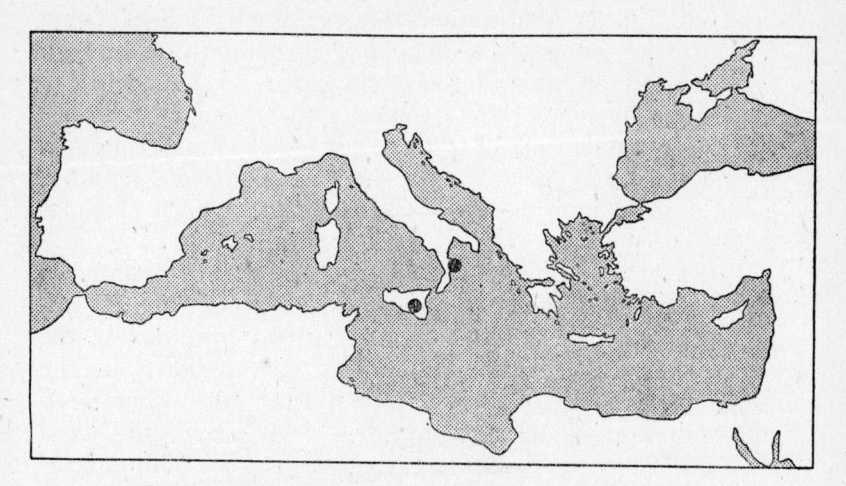

FIG. 20. SULPHUR PRODUCTION IN THE MEDITERRANEAN

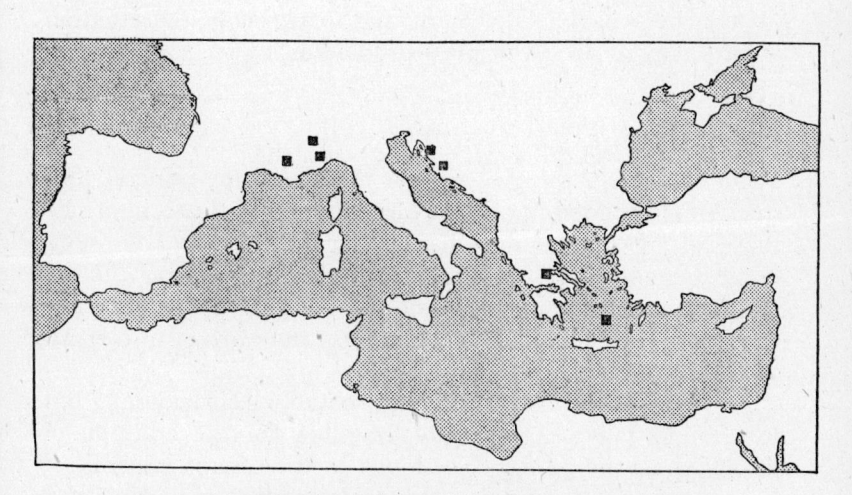

FIG. 21. BAUXITE PRODUCTION IN THE MEDITERRANEAN

the first stages of manufacture on the spot; but as coal is needed for zinc and coke for lead, it seems to be more sensible to transform these minerals elsewhere, more especially as zinc or lead mines are exhausted fairly soon. To build a factory alongside a mine which is liable to have a short life means running the risk of being left with capital locked up in equipment that cannot be used. It is much wiser to build the foundry near the coal mines, in other words near the English Channel or the North Sea. As these districts are densely populated and well endowed with financial capital, trade currents requiring shipping lines and cheap freight rates are naturally created. In this way the minerals are drawn into, almost seized upon, by this centre of attraction.

If it is decided for reasons which are not strictly economic to erect a foundry in the Mediterranean, it is not usually put close to the mine. It is more likely to be situated at a seaport which is easily accessible, and where it is possible to handle the products of several mines. If one mine is exhausted it is still possible to carry on with the others, as the foundry does not depend exclusively on any of them. The fact does remain, however, that the industrial centre of gravity lies outside the Mediterranean area. It is north-west Europe with its resources of coal and iron, with its great accumulations of men and capital, with its organised markets and its enormous buying power, that is the magnet.

Such is the situation in the Mediterranean. It accounts for the currents which are easily discerned in the exchange of the raw materials which are exported or imported as required.

The Mediterranean is a chronic importer of coal to the extent of from 16 to 18 million tons a year. Italy imports 14,150,000 tons, Egypt 1,575,000, Greece 810,000 and Algeria 675,000, quite apart from the requirements of Jugoslavia, Tunisia, Morocco, Syria, Palestine, Malta and the Mediterranean coast of France. If we include a very small export of a little more than a million tons a year from Turkey, and a production of about 16 million tons for the whole of the Mediter-

MINERAL RESOURCES

ranean, we arrive at a consumption of 33 million tons, corresponding to only 2.65 per cent of the world's consumption. This is very little, indeed almost nothing, if we compare it with the consumption of the United States at 400 to 500 million tons and

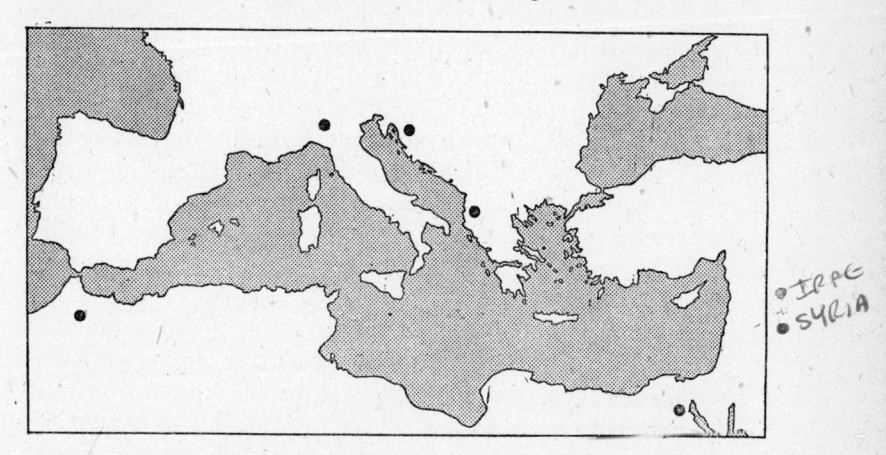

FIG. 22. PETROL PRODUCTION IN THE MEDITERRANEAN

England with about 200 millions. No doubt the Mediterranean requires little coal because of its climate, but the lack of demand is chiefly because it is not an industrial region. Also, it is not industrial because there is no coal. The economic consequences thus are decisive.

The principal Mediterranean countries, realising this situation, have made vigorous efforts to free themselves to some extent at least from their dependence on imports from England, Germany and other northern districts. The development of water power in Italy and France has meant a certain amount of progress, but it is still very limited. Harnessing waterfalls and the use of petrol provide a little alternative power, but in the present epoch coal remains a fuel which a civilised country simply must have. In these circumstances the Mediterranean labours under a grave handicap, and although it may be attenuated somewhat it is still very serious.

As for the minerals, in this case commerce follows an entirely different course. Iron ore and chromium are exported in the raw state towards north-west Europe to be handled in the great centres of coal and metallurgy there. Italy is an importer of iron ore and scrap iron, but she is the exception. She has always wanted to possess an iron and steel industry, although she has neither coal nor iron of her own in any quantity.

There are of course a few stray local foundries in the Mediterranean — lead foundries at Megrine in Tunisia, at Cartagena and Malaga in Spain, at Estaque in France, at Pertusola in Italy, at Laurium in Greece, and at Balia in Turkey; and zinc foundries at Crotone and Mestre in Italy; and a brass foundry at Bor in Jugoslavia. Otherwise lead, zinc and copper are generally exported raw or at the first stage of manufacture to the industrial countries of north-western Europe. Belgium, for example, is a great importer of zinc. Bauxite used to go in large quantities to Germany; potassium is exported to Holland, England and the United States; while phosphates and mercury are sent all over the world.

IV

These resources, as we have seen, are not negligible, in fact they are sufficient to make the Mediterranean one of the interesting mineral regions of the globe. However, it must be admitted that the total is seriously reduced, if we deduct the minerals, which, although they come from Mediterranean countries, still do not belong strictly to the Mediterranean region. Such is the case, for example, if we subtract the Turkish mines which are situated on the coast of the Black Sea or on the plateau of Anatolia; or the mines which belong to the western or northern section of Spain; or such French mines as are in the south-west or on the central plains and not in the Midi; and finally such Italian mines as are situated in Lombardy and therefore must be classed with those of central Europe. Admittedly these are arbitrary classifications, but even in interpreting them very liberally the Mediterranean can never claim to possess any great mineral wealth. Whenever it is a question of coal, we come back at every turn to the discouraging refrain: lack of coal for fuel.

It will now be in order to compare the situation in ancient times with that of the present day. In the Greek or Roman period non-ferrous metals were a step in advance of iron, for this civilisation was based more on copper, zinc, lead and brass. These metals were sufficient for the needs of an epoch which was scarcely industrialised, although for that matter the Mediterranean was always well off in this respect. The ancients possessed mines at Laurium, at Cartagena in Spain and in North Africa. The tremendous demand for raw materials which characterises the trade of the present time did not exist. The centre of gravity of economic life lay elsewhere. It was not a period of heavy metals and masses of cheap merchandise; what they wanted was exactly the opposite. Consequently the civilisations situated on the borders of the Mediterranean did not suffer in any way owing to the limitations which assume so much importance today. They were able to hold their place and even dominate the world without the slightest difficulty, for the essentials of their time were not lacking.

We are, however, now at the age of metallurgy, and conditions have changed completely. The economic period which began with the eighteenth century produced a feverish industrialisation which became synonymous with power. Under this regime, coal by its geographical position determines the topography of the wealthy countries, which are at the same time powerful politically. When we take a bird's eye view of the Hercynian folds, the entire industrial belt of the Western world stands out in relief. This belt begins in the Alleghenies in America, and continues as far as the Donetz Basin and the Ural Mountains, passing by way of England, northern France and Belgium the Black Forest, and includes Bohemia and Silesia. . . .

As the Mediterranean lies outside the Hercynian folds, the magnificent expansion of the eighteenth and nineteenth centuries has taken place almost entirely outside its boundaries. The great inventions that have turned the world upside down have been conceived elsewhere. It is elsewhere that heavy industry, which is typical of the modern world, is situated. The dense centres of population, of capital and buying power, are concentrated elsewhere. The Mediterranean in so far as its mines are concerned is simply an exporter of minerals, like the new countries and colonies.

These considerations dominate not only the problem of minerals but the industrial problem as a whole. Up to the present the Mediterranean has hardly been industrialised at all. If existing conditions were modified for any reason, it could in its turn alter its course towards industrial development. Certain indications suggest that such a change might be possible, but it would mean a radical alteration in the equilibrium of the continents.

INDUSTRIAL DEVELOPMENT

tural raw materials the position is just as favourable. Wool amounts to about 9 per cent of world production, silk to 9 per cent, cotton to 6 per cent, cork is plentiful, and so on. The forests, of course, are mediocre, and there are no tropical products. But the foodstuffs which come under the term 'Mediterranean' are on the contrary a marvellous source of wealth: citrus and other fruits, olives, wines, etc., are grown in abundance. There is a certain amount of wheat, and the barley amounts to one-fifth of the output of the whole world; but there is little meat and not a great deal of fish. As a result the Mediterranean is obliged to depend on its imports of basic foodstuffs, and although luxuries are to be had in quantity this cannot be said of any of the necessities.

Work people are numerous, and on the whole do very well. Italian manpower is well known for its excellence, and the Spanish agricultural worker is vigorous and energetic. The African labour centres — Sousse in Morocco and Kabylie in Algiers — also are important sources of manpower at wages which have been low enough for many years. These people have a certain traditional skill, and in addition they can be uprooted easily. The craftsman of North Africa perpetuates an earlier economic period and this is not without value, for although his tools are out of date he has a real technique. He will go on working for a poor remuneration as he combines work and religion, and the result is social peace and a contented atmosphere.

Once attracted into the factories, however, these men lose many of their fine qualities. Their output is poor, and they do not like working with machines. They are not interested in their pay envelope, and when they have satisfied their modest requirements they consider that they have made enough effort and they stop working. It is just the same on the European shore of the Mediterranean, where factory hands also are not entirely exempt from these economic shortcomings. They rarely have the same appreciation of persistent effort that we find in the north and in the United States.

THE MEDITERRANEAN

When it comes to the purchasing power of the Mediterranean people, we again encounter drawbacks. Even in Europe, but especially in Africa, frugality is the dominant note. Needs are so few in these parts, that expenditure is incurred as much for vanity as for anything else. Here let me emphasise the archaic character of the purchases of the native population in the Mohammedan countries. Owing to the diversity of races and religions the demand is extremely varied and standardisation is very difficult, for the Jew, the Arab, the Berber and the Fellah each has his own customs, needs and equipment. What a contrast with the American customer, who passes through the rolling mills of assimilation and comes out all ready to absorb the results of mass production!

In North Africa, however, colonisation by the white race has had a rejuvenating effect on purchasing power. By their very presence the colonials have developed an important demand for modern merchandise. This may be for building materials such as lime, cement, tiles, bricks, etc., or for the production of foodstuffs, for which fertilisers, agricultural implements and tools are needed. Or again they want modern forms of transport like motor cars and aeroplanes, and then too they must have amusements. This trend is very marked in Algeria, Tunisia, Morocco and Egypt. In any case if we compare the capacity of the Mediterranean market with that of North America or the British dominions, the advantage certainly lies with the new communities overseas.

The Mediterranean is a good field for the formation of capital. It accumulates spontaneously by means of savings, for owing to their very modest needs the people spend next to nothing. At the same time large commercial fortunes are built up at Marseilles, and in Greece and Egypt, and large agricultural fortunes in Algeria, Tunisia and Morocco. And yet even these fortunes are of limited proportions, for capital does not accumulate nearly so quickly as in north-west Europe or the United States. What do these people do with their money? Traditionally they like to buy more land, but in the case of

INDUSTRIAL DEVELOPMENT

large estates which have been built up rapidly, money is frequently invested in industrial or mining shares. A successful man is ready to risk some of his profits, and especially in local enterprises, as this gives him an opportunity to show his patriotism, his devotion to his city, to his province, or to his country large or small. Vanity is not always absent from such investments which flatter the newly rich.

Whatever these capital resources may amount to, they are not sufficient to serve as the basis of any great industrial development. It is necessary to go further afield whenever really important sums are needed; thus Egyptian industry and the mines in Spain and Africa have been financed with foreign capital. In the French Midi local financial contributions are notable, but the large undertakings as they have steadily expanded have had to look for money elsewhere. One may say the same of Italy, indeed generally speaking the financial centre of gravity of the Mediterranean lies outside of the region.

The question is, do we find that spirit of industrial progress which as I have already said is one of the essential factors of the industrialisation of a country? The Mediterranean possesses an incontestable tradition of craftsmanship and gifted workmen, but by inclination and custom, it is commercial rather than industrial. Furthermore, prevailing economic conditions do not provide any incentive towards progress, either in the direction of research or of large-scale mechanical production. The exceptions to this are the French Midi and Italy where labour is remarkably skilled mechanically, and in a general way not expensive enough to make further mechanisation an urgent matter. Yet the driving force which is in the very atmosphere of Manchester, Lille, Essen or Detroit does not exist beneath these all too happy skies.

Mediterranean industry is thus limited to a certain kind of manufacture. Local or imported raw materials receive the first stages of conversion here; and certain articles of consumption are produced from regional products, or to satisfy the needs of the local people. Various minerals are washed and sorted, although even these stages in their treatment are definitely limited — as we know! — by lack of coal. Next come the chemical industries based on the local supplies of sulphur and phosphates. Then we have numerous building materials such as lime and cement, tiles, bricks, etc.; various industries connected with foodstuffs such as flour milling and the manufacture of semolina, biscuits, rice products, preserved and conditioned fruits and vegetables, sweets, wines, oil, sugar, beer, distilleries, etc. Imported colonial products, notably vegetable oils, are treated here; various textile operations are carried out such as the spinning and weaving of cotton, wool and silk; while rope and twine, jute sacks, packing materials and colonial equipment also are manufactured. There is a certain amount of mechanical and naval construction, either in the form of heavy industry, or more often of craftsman's repairs or other small maritime equipment; at Marseilles, certain specialities such as ship screws and propellers are manufactured.

We soon grasp the nature of this industry. In the main it is a question of consumer products rather than equipment, for we know that it is not, and cannot be, metallurgical. Foodstuffs, when it is a case of the so-called Mediterranean products, naturally tend towards quality, almost to luxury; but when it comes to the treatment of raw materials, that has to be undertaken by heavy industry, so any manufacturing of this kind in the Mediterranean is quite exceptional.

It is easy enough to discern the laws which have dictated the geographical situation of industry in the Mediterranean. As is to be expected it is chiefly located in the big ports, more especially as these usually are large cities to which coal and raw materials can easily be imported and which afford plenty of manpower. From these maritime centres merchandise can readily be distributed to local markets or exported to foreign countries. Marseilles, Genoa and Barcelona, where production is determined by commerce and the sea is close at hand or not far away, are industrial cities of this kind.

There is still another group of a different type that is not

INDUSTRIAL DEVELOPMENT

genuinely Mediterranean. In these cases industry is situated near fuel supplies — when there are any — at the source of some raw material, or in the midst of dense centres of population. The industrial region of Milan, the Egyptian sugar industry, and the metallurgical industry of Alès all fall into this classification. We must bear in mind, however, that we are considering territories now that strictly speaking belong to the continent rather than to the Mediterranean. In fact, except perhaps at Milan, there are no great industrial regions here, nothing like the Ruhr, Lancashire or Flanders. Industry is always sporadic and definitely of secondary importance. This is just the nature of things.

III

In all this we must not be too sure of ourselves, for new circumstances are already modifying industrial conditions everywhere.

Economic factors of production no longer follow the same tradition as they did yesterday and the day before. Energy derived from water power and petrol can now be used without recourse to coal. Furthermore, the latest types of machinery make it less necessary to depend on technical skill, in fact anyone at all, even someone quite foreign to the work, can be taught all that is required. Distribution on the other hand is becoming an essential factor, and therefore in an increasing number of cases it is definitely an advantage to be near a consumers' market; this in turn is leading to an appreciable amount of decentralisation. Finally, great enterprises, whether national or international, are now inclined to look upon a factory simply as part of the whole, according to whether their plans are for regional or continental decentralisation. In this way the laws of industrial topography give way to a planned economy which no longer respects the individuality of a factory, and may even reduce it to the rank of a workshop.

Alongside these factors which are genuinely economic, an increasing number of other factors which cannot be described as economic at all are pressing for the industrialisation of regions which frankly are not suitable. Now that nationalism has become the overriding argument, certain countries are anxious to possess industries and maintain them artificially for political reasons, even when these industries are unjustified economically. Work must be provided for the people, it is argued, local technicians who have been imprudently trained in too great numbers must be kept in employment, and there is also the desire to imitate the great industrial nations! To this must be added the rapid accumulation on all sides of great fortunes, which are looking for semi-industrial or semi-speculative investments, and here the spirit of vainglory is by no means excluded. Perhaps, too, the new countries are jealous of the old, whose success they wish to emulate in their turn, and they also are wondering if they are ever to come of age economically.

Reasons of this kind, economic and otherwise, are dragging the Mediterranean like so many other regions into the industrial movement. Although lacking in both iron and coal, Italy insists on having a metallurgical industry; and Turkey, in order to assert her independence, also has her industrial programme mapped out. French North Africa is particularly interesting in this respect. Raw materials on the spot are asking to be converted; manpower is abundant and cheap; the purchasing power of this young country is already expanding rapidly; large fortunes are accumulating as a result of youthful economic activity and enthusiasm; patriotism towards continental Africa, and finally the insecurity of Europe — such are the arguments put forward inviting people to invest in the industry of Algeria, Tunisia and Morocco. The manufacturing expansion of Algeria in the past few years is most striking, and the same applies to Morocco. One need only make a cursory survey to realise the flourishing point to which certain industries have recently developed, notably building, chemical products, machinery, various types of textiles, breweries, tinned foods and the local manufacture of foodstuffs.

INDUSTRIAL DEVELOPMENT

We find the same recent expansion when we turn to Egypt, where Alexandria is becoming a large industrial city. Egypt produces cotton, sugar, a variety of foodstuffs, and possesses immense fortunes, a vast working force that can be hired for almost nothing, and a strong consumers' market. The Egyptian Government at the same time is eager to make itself felt. With

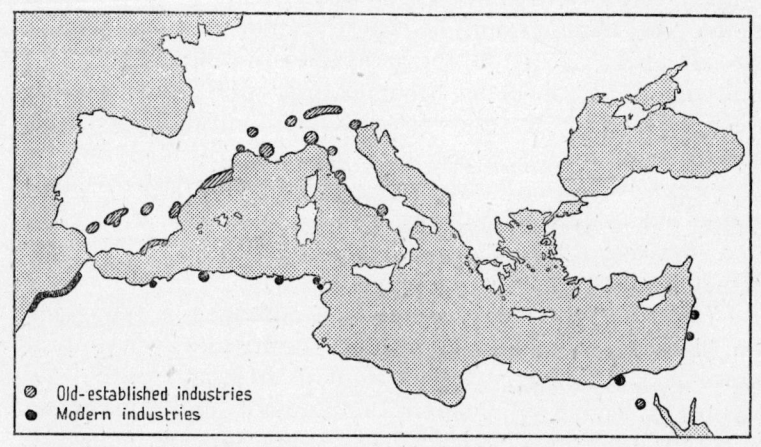

FIG. 23. INDUSTRIAL CENTRES IN THE MEDITERRANEAN REGION

the Bank of Misr acting as intermediary it has created and now maintains and supports various industries. These industries could have been established by private initiative, but they might not have been able to stay the course. Similarly we should note the beginning of industrialisation in Syria and Palestine (Fig. 23).

How far is this policy of industrialisation justified economically? I should say only so far as it responds either to existing circumstances, or to the Mediterranean's own particular advantages. In view of the world revolution and the decentralisation of industry everywhere, and also taking into account the growing insecurity of Europe, it is natural that the North African people should wish to build factories of their own. But this development becomes unreasonable if they insist on sheltering behind a protective tariff or political support, undertakings which countries specialising in heavy industry would do better. Obviously there is no sense in creating heavy industries where there is no coal, or to try to manufacture locally things which one can buy much more cheaply elsewhere.

It is imprudent on the other hand for a country which exports raw materials to try to develop an industry of its own at the same time. North Africa being an exporter of wheat and wine, and Egypt an exporter of raw cotton are well aware of this; and South America will yet have to learn this lesson at her own expense. In this connection the English have a wise proverb: 'You can't sell the cow and have her milk too!' One may say no doubt that a newly created local industry, by the very fact that it absorbs part, regulates the market for the raw materials that it exports. One can also say that the excess population must be given work. Nevertheless in a country of this type, it is the exports of raw materials which today as yesterday are the essential factor in the balance sheet. Now these exports have as their opposite numbers the imports of manufactured articles. It might be dangerous to offend the foreign customer, or even to diminish his buying power, by undertaking to manufacture the things which he is already manufacturing. In addition industries in new countries can work only for their own home market, but there a certain price ceiling is soon reached. After that the profit, which looked magnificent at first, suddenly falls away.

The outcome of this discussion is that to replace a trading system by one that is entirely new is a delicate matter. In the Mediterranean there are certain regions that belonged to the old economic civilisation, notably the French Midi, Catalonia and northern Italy, where industrialisation has come about normally, principally because they are part of the life of the continent. But on the whole the Mediterranean remains central and northern Europe's source of supply for raw and semi-manufactured products which are received in exchange for finished articles. Such is the basic and complementary

INDUSTRIAL DEVELOPMENT

nature of the trade between the two zones lying to the south and to the north. To tell the truth the Mediterranean is not really industrial, either in its transportation or its economic climate. Its individuality does permit a limited amount of industry, but any further development risks going beyond the bounds of its particular genius.

One may say the same thing of many other countries outside Europe which at the present time are on the road to industrialisation. Are they justified in taking this course, and will they not regret it in the end? The economic wisdom of yesterday may have thought so, but the wisdom of tomorrow may see matters from another angle. The whole world is upside down, and we wonder if we should not say with the witches in *Macbeth*, 'Fair is foul and foul is fair!'

CHAPTER XII

THE MEDITERRANEAN PORTS

It is impossible to study the Mediterranean ports without speaking of the ancients, either to recall them as forerunners, or to emphasise the enormous changes which have taken place since their day.

In ancient times there was a great contrast between the moderate value of the ships, and the high market prices of the merchandise which these same ships transported. In such conditions the ship had to accommodate itself to the port, but in modern times the port is built to fit the ships. In addition, the problem of communications inland was of secondary importance, for the products though costly were light in weight and easily carried no matter how bad the roads were. This does not apply to the heavy raw materials of the present day. Economically we are considering two different worlds. The solutions to the problems which were faced by the ancients — the Phoenicians, Greeks, Romans, Carthaginians and Byzantines — were so sound that we still make use of them. The ancients blazed the trail. But the problems of our time are of an entirely different order.

In the beginning the main consideration was the security of the port, either against the sea or against pirates. These early ports consisted of a grounding beach where the boat could be moored at the foot of a fortification, within which men and merchandise could be sheltered. There was also an intermediary space corresponding to the quay of the present day. In his excellent book, entitled *Homme et la Côte* (Man and the Coast) published by the Nouvelle Revue Française, M. Marcel Hérubel singles out the ports of Pharos, Phocaea, Tyr, Sidon and Piraeus as special prototypes of the Mediterranean thalassocracies (Fig. 24, p. 165.) Changes came with progress, and

THE MEDITERRANEAN PORTS

jetties were built, also piers, dykes, docks and platforms. The port of Pharos, for instance, had five miles of dykes, and 300 acres of docks. Carthage had its docks and its sea wall (Fig. 25, p. 166). The Romans and Byzantines perfected this technique still further, and later on the Genoese transmitted the

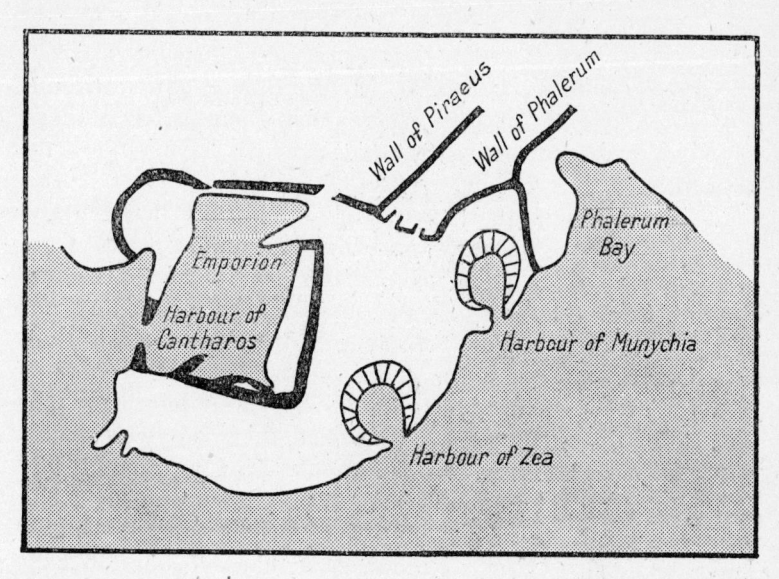

FIG. 24. THE HARBOURS OF PIRAEUS IN ANCIENT TIMES, ACCORDING TO HÉRUBEL

tradition to the West. Another change occurred in the Middle Ages, when the bazaar developed independently of the seaport.

In ancient times the ports were generally situated in indentations along the coast, at the base of cliffs, in little bays with sheltered beaches, in rocky coves, or on an island which was then artificially linked up with the mainland. The ancients always tried to protect their harbours from flood water, but when they chose a spot near the mouth of a river they were not always successful.

Both Thucydides and Vitruvius analysed conditions in their day with great perspicacity. The Phoenicians, according to Thucydides, established their ports 'on bold headlands and neighbouring islets'. In a page which has now become a classic, the Greek historian adds that 'the maritime establishments on the shore were surrounded with walls. People took possession of isthmuses for the benefit of their commerce, and also to fortify themselves better against their neighbours'. According to Vitruvius the natural harbours were installed in bays formed between two rocky points, or between two small promontories jutting out to meet each other. In the enclosure they built porticos, docks for their boats, warehouses, and towers from which chains could be stretched across from one to the other. The West, learned these ancient traditions but later abandoned them. However, we have now gone back to them again, and are following them so closely that conditions do not seem to have altered at all.

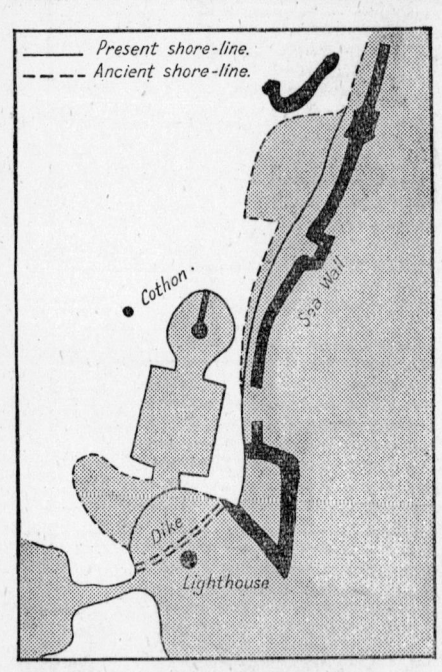

FIG. 25. THE ANCIENT HARBOUR OF CARTHAGE, ACCORDING TO HÉRUBEL

In 1828 Hérubel and Dulau, in the course of their maritime work, analysed as follows the conditions required for the establishment of a good harbour in the Mediterranean: 'Choose a cove with no river emptying into it, and make sure that there is sufficient depth of water to keep the shore current from

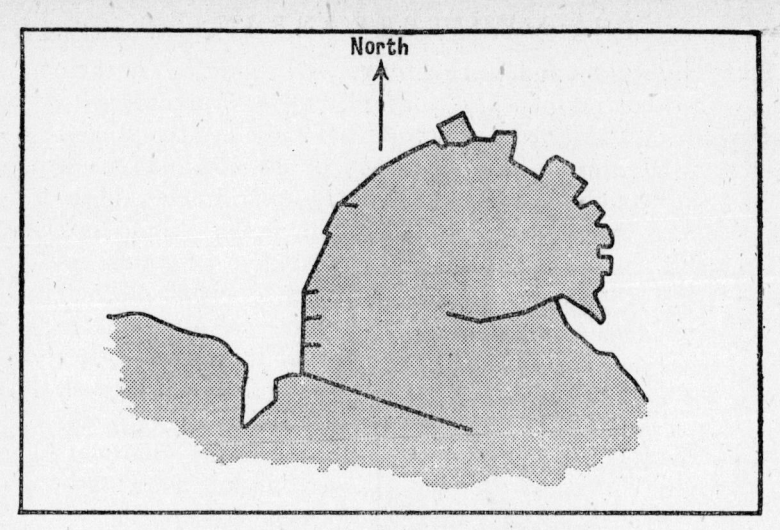

FIG. 26. THE PORT OF GENOA IN 1875

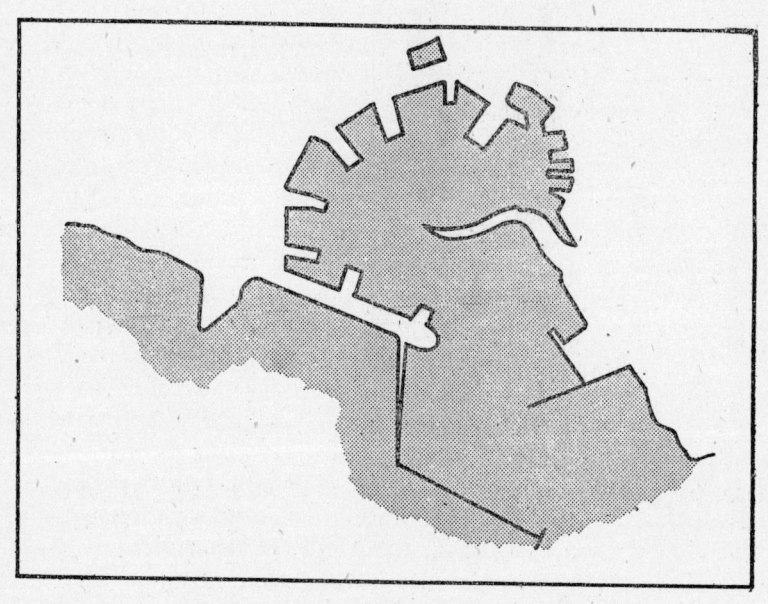

FIG. 27. THE PORT OF GENOA IN 1890

throwing out either sand or silt when it passes between the two points of land closing in the entrance to the cove.'

Entirely new conditions arose, however, at the beginning of the nineteenth century. The Atlantic tides presented new problems, for the ships' tonnage had increased and so had the weight of the merchandise transported. Consequently the wharf, sheds and warehouses, and the outlet to the interior of the country became of first importance. In this respect the Mediterranean was handicapped. In ancient times it had been well situated owing to the numerous bays along its coasts and the absence of tides, but the moment that heavy loads had to be transported inland it was less well off.

Nevertheless, in matters pertaining to the sea the Mediterraneans have always been pioneers. They faced the initial problems of navigation and solved them, they invented the Latin sail and the stationary rudder, they cut a canal through the first isthmus, built the first lighthouse, and wrote the first book describing seaports and giving their situation. And speaking of seamen, Conrad was not far wrong when he said that the Mediterranean had 'cradled the profession in its infancy'.

II

Technically the establishment of the Mediterranean ports has been determined by the geographical structure of the region, which in its turn is the result of the Alpine formation. The shore appears abruptly, after which the sea floor descends to great depths. The swiftly-flowing rivers form, not estuaries, but deltas with dangerous alluvions which shift with the currents that come in from the sides. Furthermore, the vertical rampart of mountains around the coast makes it difficult to reach the interior. Such are the obstacles. On the other hand the lack of tides is an obvious advantage, for boats can come in and out of harbour at will at any hour without having to worry about the depth of the water.

THE MEDITERRANEAN PORTS

The situation of the various Mediterranean ports may be divided into two categories that are entirely different: first, deep-water harbours where the coast is rocky; and second, those on the lowlands.

The ports on rocky coasts are usually situated in small bays which provide sufficient shelter for small craft. The old harbour of Marseilles, the Vieux Port, is a case in point, and another

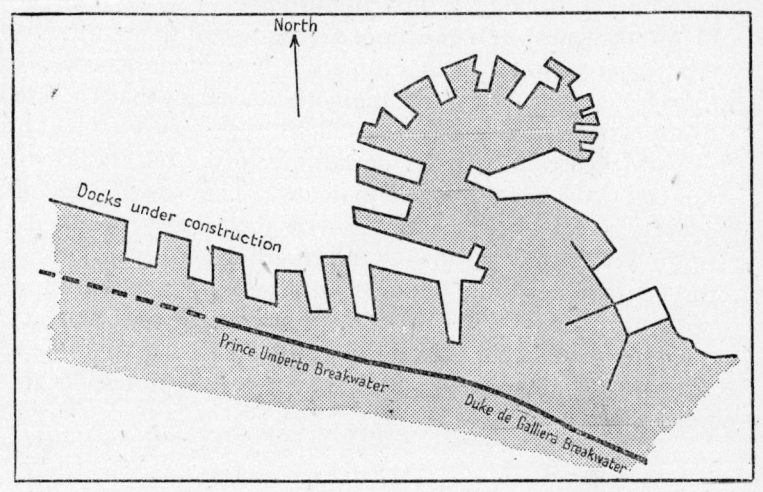

FIG. 28. THE PORT OF GENOA IN 1929

is the old port of Genoa (Fig. 26, p. 167). The development of this type of port soon encounters serious obstacles, for as a rule there is a range of mountains in the vicinity and this makes it impossible to extend further harbour construction inland. Progress is blocked in this direction, so development must take place to the right and to the left, between the shore and the mountains. As the deep water is almost always limited, the harbours are carried far along the coast. This is what has happened at both Marseilles and Genoa.

A third phase consists in reaching the sea logically by building dykes to provide protection, and these eventually

THE MEDITERRANEAN

lead to the construction of new docks. Here again Genoa and Marseilles are typical examples (Figs. 27 and 28, pp. 167, 169). At Marseilles one can trace the same phases of evolution as at Genoa: the first little bay which is now the old harbour; a series of docks built close to the sea — Joliette, the National Dock, de la Pinède, and President Wilson (Fig. 29); finally there was the possibility that nature refused at Genoa, i.e. the development of a second harbour at Caronte on a low-lying stretch of coast (Fig. 30, p. 171). As a result the port of Marseilles virtually extends from the Bay of Marseilles to the mouth of the River Rhône, and thus its harbours come under both categories, rocky coast and lowlands (Fig. 31, p. 171).

One can easily grasp the advantages and disadvantages of these deep-water ports on the Mediterranean. There is no need to dredge or calculate the depth of the dock, but on the other hand it is difficult to build dykes as the foundation gives way almost immediately. Handling the cargoes constitutes a particularly thorny question owing to the lack of sheds, while transportation inland discloses railway problems that are very difficult to overcome. The deep-water harbours at Marseilles, Genoa, Naples, Beirut, Constantinople,

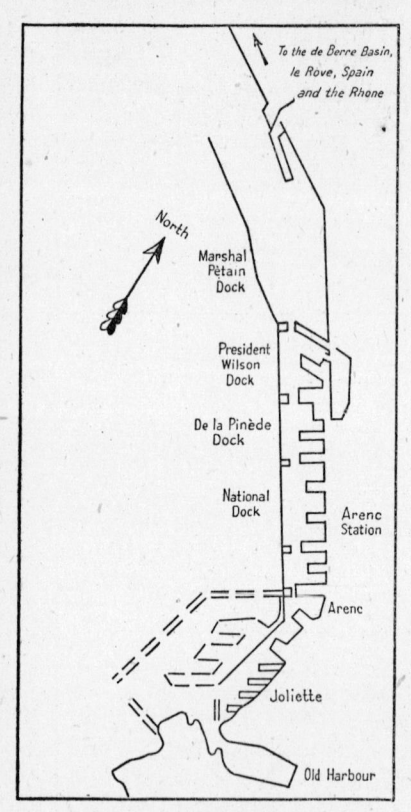

FIG. 29. THE PORT OF MARSEILLES

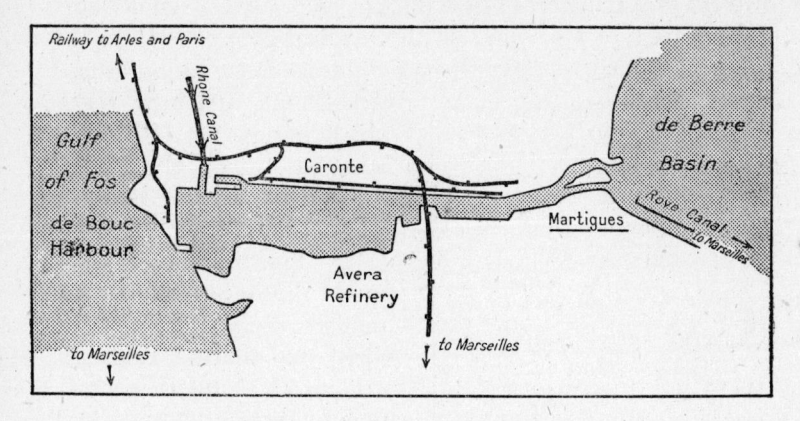

FIG. 30.   THE PORT OF CARONTE

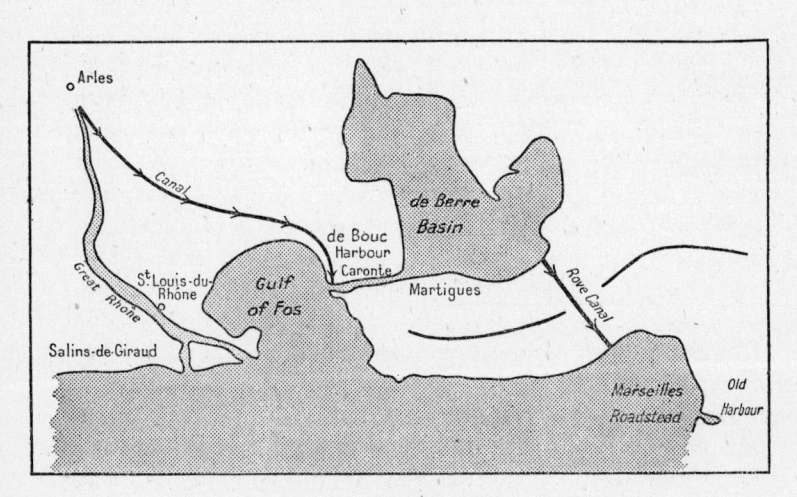

FIG. 31.   THE PORT DISTRICT OF MARSEILLES

THE MEDITERRANEAN

Algiers and Oran are the most typical of the Mediterranean.

Now the ports situated on the lowlands, on the other hand, have plenty of docks and no difficulty whatever in reaching the interior of the country. But they lack depth, and are shallower still if sand drifts in from the mouth of some neighbouring river or delta. For this reason it is essential that they should be built facing the direction opposite to the current. Alexandria, for example, is protected from the alluvions of the Nile which the current carries towards the east, but Port Said is not. Venice, Sète, Sfax and Bizerta are typical low-lying ports, usually with docks like those on the Atlantic coast, the English Channel or the North Sea. They are very different in structure from the ports of Marseilles and Algiers, which one might describe as maritime cities belonging to a different family (Fig. 32, p. 173).

III

The functions of the Mediterranean ports are varied and complex. Some ports are quite simple, but others give a variety of services. It will be as well to classify them in order to distinguish one from another. Thus we have complete ports which supply all kinds of services, regional industrial ports, international ports of transit, ports of call, express ports for passengers and perishable goods, ports that specialise, and finally entrepôt ports for goods in bond. Let us now try to fit the principal Mediterranean ports into these different classifications.

All these functions are to be had in the complete ports. Marseilles and Genoa are establishments of this type, which, however, occur less often in the Mediterranean than on the north-west coast of Europe. Their activities include the transport and distribution of imported and exported goods (notably the importation of raw materials for the local industry, or transit to the countries of central Europe); the reception and provisioning of ships, the transport of international passengers,

THE MEDITERRANEAN PORTS

and also various entrepôt operations. These ports offer such a variety of services that the list seems almost endless. Alexandria and Constantinople could be complete ports, and yet they are

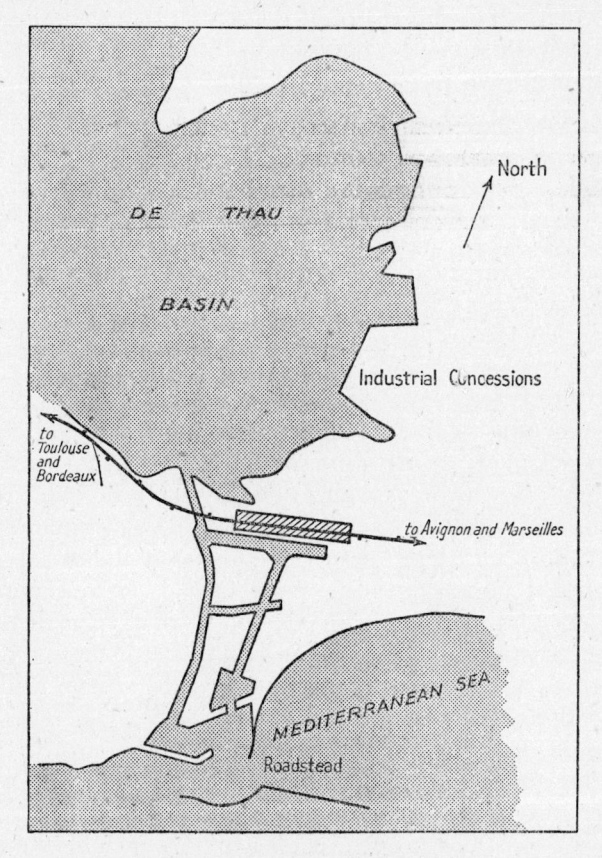

FIG. 32. THE PORT OF SÈTE

not — but I shall explain this later. This leaves the two rival ports, Marseilles and Genoa, in a class by themselves.

The regional and industrial ports undertake to export the products of a certain region, and also to distribute locally the

THE MEDITERRANEAN

goods that each territory requires. Eventually facilities for the importation of coal or raw materials, plentiful manpower and a favourable geographical position lead naturally to industrial development. Marseilles and Genoa again are ports of this kind, being great industrial cities as well as great maritime establishments.

Barcelona is both an industrial port and a port of distribution, but its activities in matters of transit, port of call and passengers do not amount to much. Napels receives passengers and undertakes to distribute merchandise throughout the region, but it has neither transit facilities nor the traffic of heavy industry. Algiers is a great port of call and of regional distribution, but without facilities for international transit, and the same applies to a lesser extent to Oran. Venice is a great port of distribution and also an industrial port, being the outlet for the whole of the Po Valley. Trieste undertakes both distribution and transit; Alexandria distributes, acts as a port of call and handles goods in bond; Salonika, distribution and transit; while Athens confines itself mainly to distribution.

The ports of transit provide a means of transport into the great country inland: it is not a question here of regional penetration or distribution, but of international penetration and distribution. We have already pointed out several times the weakness of the Mediterranean in this respect owing to its lack of large rivers and the mountain barriers along its European coast. I am not overlooking the Rhône, but there is no Danube here, and no Rhine. There are of course many passes through the Alps such as the Cenis, Simplon, Gothard, Brenner and Tarvisio, but even these are limited, and their steep slopes require great feats of engineering. Transit from the Mediterranean to central Europe is thus seriously handicapped compared with the means of penetrating into the interior from the English Channel or the North Sea. There the great rivers change little in gradient as they traverse the lowlands. As a result the main European ports of transit are in the north rather than in the south, and it requires all his extraordinary

THE MEDITERRANEAN PORTS

energy for the Mediterranean to extend, even to maintain, his own sphere of activity.

In this respect Marseilles certainly is the most important port in the Mediterranean, as it has the Rhône as a first-class means of penetration inland. Genoa, having no river, and with only railway lines which must cross two mountains in order to reach the interior, is in a much less favourable position. Nevertheless, being very close to both Switzerland and the Lombardy Plain, it offers a means of transit with which it is hard to compete. Trieste and Salonika are both in a privileged position geographically as ports of transit, and the same is true of Constantinople. But in the two latter cases it is difficult to reach the interior, and also, as one travels eastward, the competition of the Danube route becomes more and more effective. The geographical structure, the result of the geological formations, opposes a formidable obstacle from one end of the Mediterranean to the other, or at any rate one which it is difficult to overcome completely. As a result the commercial balance of the whole of Europe is affected to a marked degree.

The function of the ports of call is the embarkation and disembarkation of passengers, and supplying fresh provisions, coal, oil, etc. For example, Port Said is the greatest port in the world for taking on coal; while nearly every boat which enters or leaves the Mediterranean stops at Gibraltar as a matter of course; Algiers, formerly a great coal port, now supplies fuel oil, and is becoming more and more a centre for passengers and tourists; Istambul is the principal port on the Black Sea. Nevertheless the ports of call are becoming less important, for as petrol replaces coal, sea voyages are longer. On the other hand as passengers and notably tourists are likely to increase in numbers, we have two tendencies working in opposite directions. But, the Mediterranean being one of the world's great highways, the port of call is playing a role that is destined to last.

The express ports are those that have been specially established to link up one or two lines of navigation with the railway, and thus transfer passengers quickly and reliably, and also perishable goods such as early fruits and vegetables which must be handled without delay. In the past Brindisi was a typical express port for passengers, as it served the post road to India. We recall how Phileas Fogg, on his tour of the world in eighty days, stopped at Brindisi on his way to Port Said and the Indian Ocean. Today the rapid transport functions of this Adriatic port have been replaced by Marseilles, and to some extent also by Genoa. Marseilles actually is one of the greatest points of passenger embarkation in the world. So far as shipment of perishable goods by express is concerned, this service has become extremely important since North African produce can be brought speedily in this way to Paris and other large European cities. Port Vendres, for example, has been specially established to provide rapid communication between France and Algiers and Oran, but in this same connection we must also mention Marseilles, Almeria, Valencia, and Palerma, and in Algeria Philippeville and several other ports east of Algiers.

We classify as specialised ports those which are equipped to handle one particular product for which they are suitably adapted. As a rule they are simply embarkation wharves, with little or no provision for imports. For example, phosphates are shipped from Sfax, petrol from Tripoli in Syria and also from Haifa, wines from Mostaganem, pyrites from Morphou Bay in Cyprus, gypsum from Ancona and Port Empedocle, wines from Sète, and minerals from various other ports. Among the ports that specialise in importing coal are Marseilles, Genoa, Leghorn, Barcelona, Venice, Piraeus, Salonika, Algiers, Oran and Port Said.

When we come to the entrepôt ports we must distinguish between the transfer of goods in bond, and an entrepôt market in the full sense of the term. In the case of transit pure and simple, the goods cross the port without stopping. But in the case of transit with unloading, this operation is only completed when the goods have been loaded again on to the railway. In other cases the goods are not only unloaded and loaded again, but they remain temporarily in a warehouse, and possibly they are conditioned as well. Where transit comprises final marketing, the products are held in a warehouse, conditioned and sold, often for future payment, so that eventually there is a change of ownership. The genuine entrepôt port comprises all these activities, and notably the final sale as the crowning effort, in which case it becomes an extremely important source of wealth and activity. Marseilles which, owing to its position seems destined to become an entrepôt port, handles ground-nuts and cereals up to the point of sale. Alexandria is celebrated for its cotton market. Genoa is chiefly a coal-distributing centre for the whole of northern Italy, and consequently its entrepôt services are of secondary importance. As a matter of fact there are not many great entrepôt ports in the Mediterranean, and it will be interesting to see why.

When he built the Suez Canal, Ferdinand de Lesseps believed that important entrepôt markets would flourish in the Mediterranean zone, and act as distributing centres for Eastern products throughout the Western world. He had even intended to equip Ismailia for this purpose. However, neither Ismailia nor Port Said have developed in this way, and although the Mediterranean is the route to India and the East, it is not in the Mediterranean but rather in north-western Europe that the great commercial centres are situated. The reasons for this are that the northern industrial centres are nearer to the coal mines on which they depend, and also nearer to the zones of dense population with their vast markets and great accumulations of capital. The truth is the Mediterranean can never really compete to advantage. When we add the material difficulties of transportation to which I have already alluded, we touch one of the most serious handicaps from which the Mediterranean suffers in the present age of coal, heavy industry and mass production. This fact has an adverse influence on the normal play of trade currents in this part of the world, for it reduces considerably both the transit and entrepôt services of the ports situated on the northern coast of the Mediterranean.

THE MEDITERRANEAN

IV

To sum up, one may consider as most typically Mediterranean the ports of regional distribution, the express ports, and those specialising in certain raw materials. Not that the commercial genius of the Mediterranean lends itself less well to international transport or entrepôt operations, but at this point we are confronted by the obstacle of a geographical structure which modern technique may be able to correct but which it can never completely overcome. A mere glance at a relief or a geological map reveals without further explanation why London, Rotterdam or Hamburg rob Marseilles, Genoa and Trieste of a large part of the transportation which by rights should belong to them. Nevertheless petrol and water power on the one hand and an eventual redistribution of the industries of the world on the other, and finally the aeroplane combined with easier and speedier communications, will eventually modify these conditions. So the Mediterranean will always remain, no matter what happens, one of the great highways of the planet.

CHAPTER XIII

TRADE CURRENTS

TRADE currents between countries and continents are determined by far-reaching underlying influences. In the Mediterranean the skein is particularly complicated, but it is possible to disentangle it if the three main routes are borne in mind: (1) Trade within the Mediterranean region; (2) Exchange of Mediterranean products for those of north-western Europe; (3) Trade in transit, with the Mediterranean simply providing a passage.

Conditions are not favourable for trade between Mediterranean countries situated in the same latitude, as they all produce the same things. Instead of providing a market for one another, they are all competing in the same markets. It is always wine against wine, orange against orange, oil against oil, tobacco against tobacco, tile against tile ... No prospect here of developing any mutual trade to speak of.

On the other hand in the very nature of things a system of complementary trade has grown up between the Mediterranean region and the outside world, according to which the Mediterranean exports articles in exchange for those it does not manufacture. The industrial revolution of the nineteenth century turned north-west and central Europe into a specialised factory for the benefit of the rest of the world. As the Mediterranean played no part in this enormous effort, it was placed in the position of a colony which ships out its products and in return receives equipment and manufactured goods, to the detriment of trade within the region itself. In the past the navigation lines running east and west were busier and more prosperous than they are to-day, for now they have no real commerce to depend on.

This observation, however, applies only to the eastern and

THE MEDITERRANEAN

western Mediterranean, for the trade running north and south, between continent and continent, is exceedingly brisk. This is notably the case between France and our North African colonies. Agricultural produce and minerals are shipped to the Mother Country from Tunisia, Algeria or Morocco in exchange for French manufactures, while many shipping lines ply between ports that are perfectly equipped, even specially equipped, to handle this or that product. A similar attraction though less marked exists also between the eastern Mediterranean and eastern Europe. This commerce flows either through Trieste and Salonika, or through the Black Sea and the Danube Basin. Thus the vertical attraction is stronger than the horizontal current. When it is a question of long-distance trade, for which the Mediterranean simply provides a passage, conditions are exactly the opposite. Trade then runs east and west, deviating towards Gibraltar or the Black Sea.

A much simpler type of trade, with heavy imports of coal serving as the basis, is carried on normally between the Mediterranean and north-west Europe. For his return voyage the shipowner looks out for sufficient freight, if possible a complete cargo which will require no further handling. The tramps of 2000 to 5000 tons which brought the coal will take back minerals, phosphates or other Mediterranean products, which are needed in the northern countries just as much as the south needs their coal. The result is a traffic which is truly complementary, in fact one might compare it to arterial circulation. If one traced the coal import routes in red on a map and used blue for the exports of minerals, phosphates and sulphur, one would have a good picture of the circulation of the blood (Fig. 33, p. 181).

The most elementary type of trade consists of imports of coal, say from Cardiff to Tunis, against exports of phosphates, say from Tunis to Liverpool or any other port in northern Europe. A more complex type, triangular this time, would consist of coal exported from Cardiff to Marseilles, or more accurately to Caronte; from Caronte the boat would go east

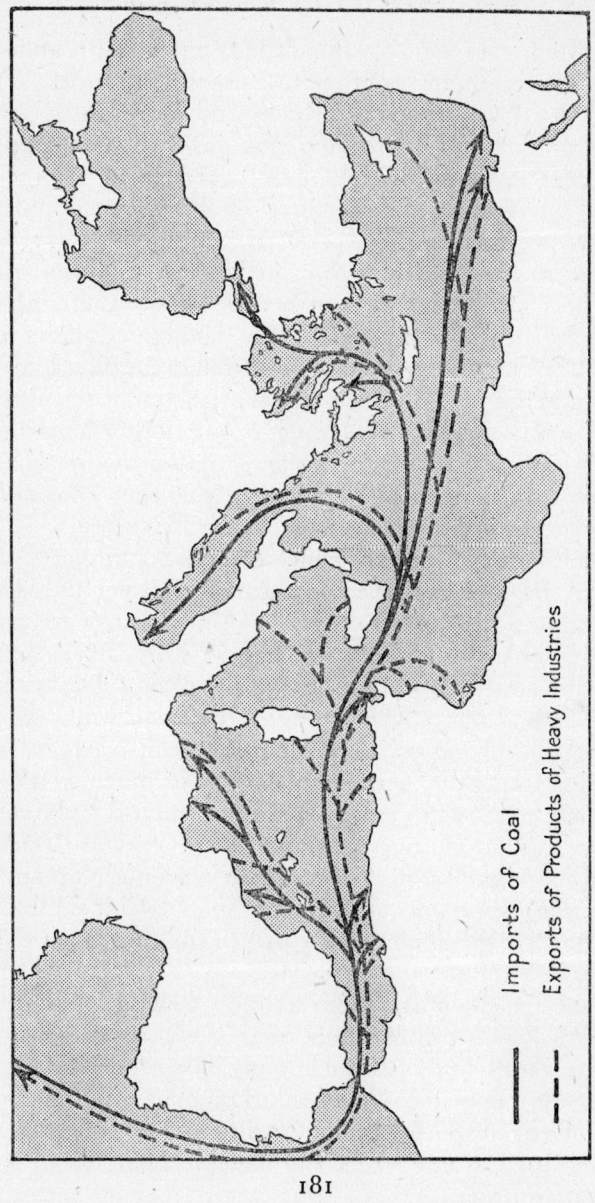

FIG. 33. COMPLEMENTARY MARITIME ROUTES (COAL AGAINST HEAVY INDUSTRIAL PRODUCTS)

——— Imports of Coal
----- Exports of Products of Heavy Industries

to St. Raphael or Toulon and load up there with bauxite to be shipped to say Rotterdam. One can conjure up endless variations of this theme, but the complementary nature of the trade always remains the same. On the other side there will invariably be coal destined for a series of ports which depend on it for a large part of their activity: in France Marseilles and Sète; in Italy Genoa, Leghorn, Naples and Venice; in Spain Barcelona; in Greece Athens and Salonika; in Egypt Port Said and Alexandria; and in Algeria Algiers, Oran and Bône.

Travelling in the opposite direction are the many heavy products which originate in the Mediterranean zone: raw cotton from Alexandria; cereals from Algiers or Oran; iron ore from Bône; copper from Salonika or Bône; lead or zinc from Cagliari, Almeria, Malaga, Cartagena, Bône, Bizerta and Tunis; phosphates from Sfax; potassium from Barcelona; sulphur from Naples, Ancona and Cyprus; chrome from Smyrna, asbestos from Limassol, salt from Tunis or Sfax; mercury from Cartagena or Almeria, bauxite from Toulon, St. Raphael or Split. . . .

Petrol would find a place on this list if it were not that tankers are an ultra-specialised type of ship, and so travel empty half the time. Apart from this, petrol traffic represents an important part of the maritime activity of the Mediterranean. There are three currents all in the same direction: (1) Russian and Rumanian petrol being exported from Batum, Novorossisk and Constanza to Italy, France, Spain, the west coast of Africa, north-west Europe, and even to Siberia via the Suez Canal and the Far East; (2) petrol from Iraq flowing through the pipe-lines of the Imperial Oil Company being exported from Tripoli, Syria and Haifa to France, England and Italy; (3) petrol from the Persian Gulf and the Dutch East Indies being shipped through the Suez Canal chiefly to England (Fig. 34, p. 183). The Mediterranean thus appears to be one of the important routes of the world for the transport of petrol.

We must place under a special heading the intercontinental

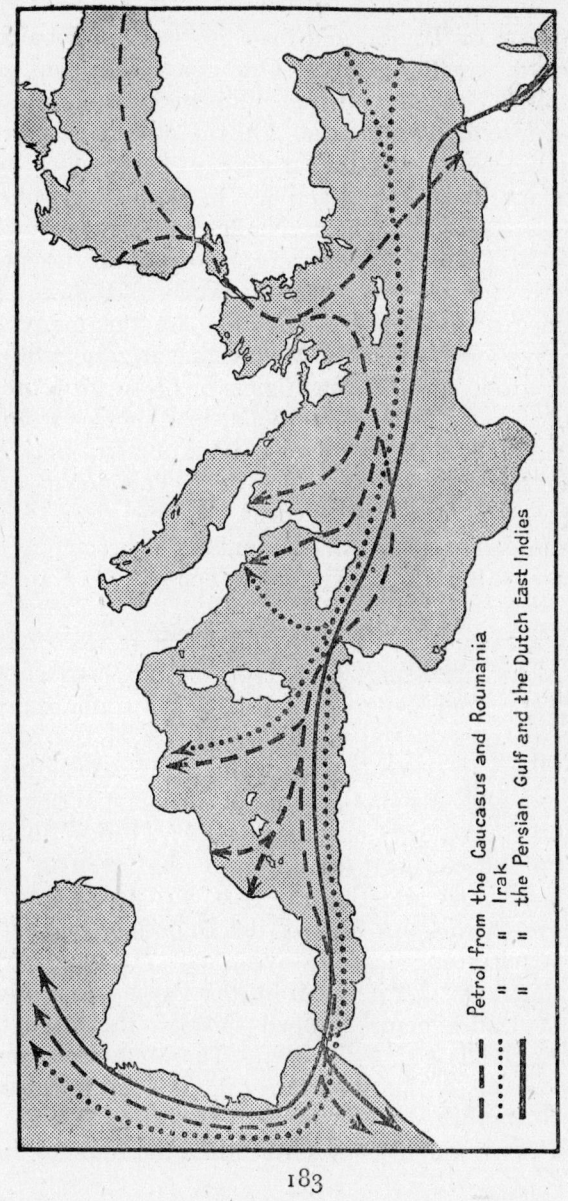

FIG. 34. OIL TRADE ROUTES THROUGH THE MEDITERRANEAN

--- Petrol from the Caucasus and Roumania
······ " " Irak
─── " " the Persian Gulf and the Dutch East Indies

THE MEDITERRANEAN

traffic carrying non-Mediterranean products to European destinations, and European products to the rest of the world. For this commerce the Mediterranean is essentially a highway, and in certain cases nothing more. Using a surgical simile, one might say that the ships coming from the Indian Ocean on their way to the Atlantic and vice versa cross the Mediterranean like a seton beneath the skin. This is the most direct route between West and East. Ships travelling to India save 42 per cent if they go through the Suez Canal instead of around the Cape, 24 per cent if bound for Japan, and 8 per cent to Australia. Furthermore, the activity of the Mediterranean benefits considerably by its ports of call — Marseilles, Genoa, Algiers, Alexandria and Port Said for passengers; Algiers, Tangiers and Gibraltar for coal or fuel oil. Here the aspect of 'passage' is the keynote.

Equally heavy and doubtless more important economically is the traffic which crosses the Mediterranean to link up continental Europe with the rest of the world. On their way in we have foodstuffs and raw materials coming from the Atlantic and the Indian Ocean past Gibraltar and Suez, to be unloaded in Mediterranean ports and sent on to the great consuming centres of the continent. On their way out we have European production, notably tools and equipment, which is distributed all over the world through such ports of embarkation as Marseilles, Genoa, Barcelona, Venice, Trieste and Salonika. This complementary trade is the result of the international economic system established in the nineteenth century, which turned the old continent into the unique and privileged provider of the industrial needs of the rest of the world. In the twentieth century this balance has become precarious and uncertain.

At this point arises the serious question of the rivalry between the Mediterranean and the ports of north-west Europe, for this transit traffic which is essential to Europe can travel equally well by way of the North Sea and the English Channel. We have already dwelt on the natural obstacles to transit via Mediterranean ports — sheer mountains through which tunnels must be cut, lack of wide rivers apart from the Rhône, and the

TRADE CURRENTS

distance of the ports from the industrial centre of gravity of the continent. Coming from America or West Africa, the estuaries in northern Europe are deep enough to permit large ships to arrive almost at the heart of the industrial region, while smaller boats even of heavy tonnage can go all the way by river without difficulty. Coming from Asia, East Africa or Australia via the Suez Canal, the argument of distance saved favours transit by Mediterranean ports; but the detour via Gibraltar and the English Channel, although it may be longer, works out cheaper in the end if the destination happens to be north of the line which divides the two economic slopes.

Europe thus has a Mediterranean slope and an Atlantic slope — or to be more precise one facing the English Channel, the North Sea and the Baltic. The dividing line naturally is changeable and indecisive, as it depends on freight rates by both land and sea, and also on the type of transport and the ingenuity with which it is used. Also the route varies with different products. For example, the economic field of Marseilles reaches Bordeaux for raisins, vanilla and salt provisions from Madagascar; reaches Mazamet for hides and wool from Australia and South America; Auvergne for rubber from Malaya and cocoa from Madagascar; includes Paris and the surrounding region for eggs and oranges from Syria, tobacco from the Levant, catgut from the Far East, and curios from China and Japan; extends even to London for certain early fruits and vegetables; to Alsace and the east of France for silks from the Far East and Egyptian cotton; southern Germany for oranges and eggs from Syria; Switzerland for cereals, raw silk and cotton; the Lyons region for silk from the Far East, Egyptian cotton, and cocoons from Turkey. . . .

Skilful handling of the route through the Rhône and adjacent canals can roll back this line, for like any other frontier it marks the balance between two opposing pressures. The struggle between the tariff rates of the two routes will always maintain an intermediary zone, like an economic beach which at one moment is covered by the tide and then left high and dry.

THE MEDITERRANEAN

Certain countries which are advantageously placed like Switzerland benefit by this competition, and they are not slow to realise it.

It is easy to see that the danger threatening the Mediterranean economy is that the traffic making the detour via Gibraltar and the northern ports may pass under the very nose of the Mediterranean without stopping. This leaves it simply as a passage — admittedly a very busy passage — where ships stop momentarily for passengers to embark and alight, or to handle costly merchandise on which the freight amounts to very little. Unfortunately this threatened danger is becoming an actual fact.

The principal entrepôt markets of Europe are not on the Mediterranean but on the North Sea. Before the war Marseilles, the largest port on the Mediterranean, handled about 10 million tons of freight annually, while Antwerp took care of 23 millions, Hamburg 25 millions and Rotterdam 46 millions! A still more striking way of showing the decline in transit is to note the low percentage of their cargoes which the Mediterranean ports ship inland by boat. At Marseilles about 500,000 tons or 5 per cent of the total of 10 million tons is carried by river, and at Genoa as there is no river the figure is exactly zero! Now at Rotterdam 41 per cent of the total freight is transported by boat, at Amsterdam 29 per cent, and at Antwerp and Hamburg 28 per cent, as all these ports are situated on rivers or beside a network of canals.

As a result of this reasoning one might be tempted to paraphrase the fable of the oak and the rose bush, and say to the Mediterraneans, 'Nature by comparison with me has indeed treated you unjustly'.

This, however, is not true of Mediterranean France, for it possesses in the Rhône a river highway capable of linking up its ports with the systems of the Rhine, the Danube or the Seine. While Italy has to scale the Alps, France has only to go round them to reach the very centre of the continent. The problem is not ambiguous at this stage of its development but essentially

TRADE CURRENTS

one of technique. From Marseilles to Sète barges of 1000 tons setting forth from the mouth of the Rhône can go up as far as Lyons on their way to Alsace or Switzerland, i.e. the Rhine, and via the Rhine to the Danube. With similar efficiency the Loire, the Yonne and the Seine are linked up with the Moselle, in other words barges can travel from central to northern France.

In this reorganisation of the French network, the limited achievements of the nineteenth century must now be revised and re-adjusted to the scale of the whole of the continent of Europe. Henceforth it will not be a question of barges of 300 or 500 tons. One must consider a minimum of 1000 tons, and up to 1500 and 2000 tons. Countries which in the light of present conditions do not reorganise their network of waterways will quickly be isolated, outstripped and disqualified. But if this programme, which is already in being, comes into full effect, nothing should prevent entire sections of central and western Europe from using the French ports as a natural shortcut in their trade with Asia, Africa, Oceania and South America.

The Lake of Constance, which tomorrow will be the key to two systems of the Rhine and the Danube, is no further from Marseilles than it is from Rotterdam. If transport conditions are brought into line, the opportunities for penetration should also be equalised. We cannot know how the conditions of world trading will develop in the coming years. It may be that in its present shattered state, Europe will cease to be the economic centre of the world. It may also be that countries outside Europe will gradually become more industrialised, and will manufacture more of their own raw materials than they did before. It may even be that economic zones will be set up that are independent of one another....

However, unless it returns to a state of barbarism, the human race cannot do without trade. The Old World will probably continue to import raw materials and export manufactures, possibly not the whole gamut of current consumption but at any rate machinery, spare parts, tools and other industrial specialities. Also, as shipping must be strictly economised during the present post-war period, the shorter maritime routes will command a premium and this is a point of great interest to the Mediterranean.

II

Now that we have analysed the trade currents, let us ascertain the contributing factors.

First of all we must consider the various types of vessel used by the different kinds of traffic. The complementary trade, which is based on coal, relies mainly on tramp steamers. Without attempting any regular itinerary or time-table, these tramps take on entire cargoes wherever they can find them. Initiative and flexibility are essential for this service, and much also depends on the individual shipowner. The service is so important that even the big shipping companies have their own tramp steamers, and they make good use of them too during rush periods.

In the Mediterranean the specialised ship is coming more and more to the fore — such boats as oil tankers, tank boats for wine, refrigerator ships and others specially built to carry early fruits and vegetables at high speed. The specialisation required to adapt speed to the commodity — the article transported — inevitably results in a lack of flexibility. The oil tanker, for example, must travel empty every second trip, and therefore unlike the tramp it is not suitable for complementary traffic. But when it is a case of rapid transport, keeping close to a timetable and establishing a regular service, the specialised boat is remarkably efficient. The boats that take early fruits and vegetables from North Africa to France are a good example, or those destined for western Europe carrying a perishable product which can be kept for a limited time only. A tank boat can take a cargo of wine to Rouen under such favourable transport conditions that it can compete successfully with a shorter itinerary by road. Obviously a perishable article, which is costly and relatively light in weight, demands a type of ship that is quite different from the one needed for heavy raw materials of lower value. The means of transport reacts in its turn, however, deflecting existing trade currents and even bringing new ones into being.

The regular steamship lines must be put in a class by themselves as they are increasing in importance every day. Certain of these companies are specially interested in the Mediterranean trade (Fraissinet, for example, or Paquet). Others link up Europe with the other continents, but have their headquarters in the Mediterranean (Messageries Maritimes, Chargeurs Réunis, Transports Maritimes, Societa di Navigazione 'Italia'). Others simply pass through the Mediterranean, possibly stopping at a port of call (P. & O., Orient Line, Norddeutscher Lloyd, Nippon Yusen Kaisha . . .), or they may take this route on their way round the world (Messageries Maritimes, Dollar Line). Finally there are the great petrol-shipping lines (Compagnie Auxiliaire de Navigation, Société Navale des Pétroles, British Tankers, Ltd.) . . . at any rate that is how matters stood at the eve of this war, and we now watch future developments with interest.

The type of ship varies from the long-distance mail boat of heavy tonnage to the rapid courier bound for North Africa, or the smaller boats of moderate speed that ply along the Mediterranean coast. The most important point to remember is that it is the regular lines owing to the very reliability of their service that actually create and establish the trade currents. One knows definitely that a ship will arrive on a certain day at a certain port, and this fact gives rise to a certain trade which otherwise would not take place. This is a repercussion in the shipping world of the standardisation that is steadily growing at the present time, and consequently the role of the regular shipping lines is also increasing in importance. The ship built by mass production is today preferred to the ship specially made by craftsmen.

The tariff of freight charges is even more influential in determining the economic currents. These rates are not based on the distance travelled as in the case of freights by rail, but on the contrary are adjusted either to the needs of the region, or to the chances of picking up a cargo for the return trip. Commercial furrows are ploughed in this way, but they depend less on physical geography and sheer distance, than on an underlying balance of complementary trade in which the outgoing current is created by the current returning in the other direction. This reveals the tremendous attraction exerted by the ports of north-west Europe as they can provide an unlimited number of cargoes, not to mention trade facilities, finance and entrepôt markets as required.

Freight charges fluctuate much as meteors crossing the sky are affected by the proximity of a star. In many instances it will not cost more — it may even cost less — to send merchandise from Algiers to Rotterdam than from Algiers to Marseilles. At the end of 1938, for example, freight rates for cereals were 95 frs. 15 from Algiers to Marseilles, and from Algiers to Rouen only 95 frs. 85; at the same date the rates for iron ore were 129 frs. and 129 frs. 70 respectively; for zinc 129 frs. from Algiers to Marseilles, but only 70 frs. from Algiers to Dunkirk! One can cite plenty of other examples of the same kind. In consequence a deviation towards Gibraltar and the Atlantic has developed at the expense of the shorter routes via the northern ports of the Mediterranean. Far from being a purely circumstantial occurrence, the reasons underlying this deviation are fundamental, being the same reasons as have made the north-west of the continent the industrial centre of gravity of Europe.

Finally, in the complex formation of this commercial network one must not overlook the existence of a purely political factor, which is perhaps more influential in the Mediterranean than anywhere else. Normally trade is a strictly commercial matter, but some of the currents which we have discussed are at least partially provoked, or accentuated, or maintained by politics. A Mother Country is naturally anxious to maintain regular communications with her colonies. This is necessary either to transport civil servants and the army, or to keep the colonies supplied with food, or simply to keep the postal service going. These communications are as much a public service as a commercial undertaking.

Before the war both France and Italy carried on this type of navigation in their relations with northern Africa. Commerce profited naturally, and occasionally new business resulted. Even if the inspiration is more political than economic, new currents which would not have existed otherwise can be created in this way, or shall we say they would be much smaller were the Government not there to encourage and maintain them. So far as relations between France and North Africa are concerned, the original political factor has been to some extent absorbed by a later development, which is genuinely commercial and quite sufficient to maintain the present traffic.

One may classify under a similar heading the prestige navigation which keeps the national flag flying on this or that coast. This kind of national publicity is planned and carried on abroad without thought of the cost of the service. Conditions in the Mediterranean, and particularly in the eastern Mediterranean, would be very different if the traffic in the pre-war years had depended on purely commercial considerations. The question now is whether preoccupations of political prestige will exist in the future. Probably they will, for we are now in a period of planned economy when politics count more than economics.

CHAPTER XIV

THE BALANCE OF TRADE

WHENEVER there is any widespread shortage of food the Mediterranean suffers more than any other region, owing to certain structural characteristics which normally are hidden. So far as it can exist at all this economy depends upon its trade relations with the other continents. These relationships, which were established in the nineteenth and twentieth centuries, are at once simple and complex, and not easily adjusted to any break in the normal routine of trade.

Let us take for example a country that is purely Mediterranean. There is no such thing of course, for neither Algeria, nor Italy, nor France, nor Egypt, nor Spain are completely Mediterranean except in theory. But let us suppose that they are, just for the sake of argument. To what extent can this theoretical country live on its own resources? What will its trade amount to, and how will it pay for its imports?

To begin with we do know that this country will possess a variety of foodstuffs — fruits and vegetables — but none of the main necessities in any quantity. No doubt it will have some meat, but very little, some wheat also, but not enough for a normal diet. There will be some rice, but neither sugar, tea nor coffee. On the other hand there will be plenty of wine. Even with the frugality which one expects from people who are naturally and traditionally sparing, they will still require a list of commodities which the changes of the last half century have made even partially necessary.

Now in the twentieth century man does not live by bread alone, but also by coal and petrol. No modern community can do without these products, unless it is willing to accept a purely rudimentary existence. The Mediterranean lacks these necessities, and the little water power that it has is not enough to take

THE BALANCE OF TRADE

their place. Thus economic self-sufficiency is out of the question, and some system of commerce is required. This is indicated by the very nature of things, for the regions that produce coal, wheat or meat also consume the very things that the Mediterranean can supply in abundance — raisins, olives, citrus fruits, nuts, figs, cork, tobacco, etc. So the Europe of the forests and fields, the black Europe of the mines and factory chimneys, is complementary to the Europe of the gardens and the sun. This is one of the fundamental elements of the equilibrium of the continent.

But can these gardens in the sun produce enough to make up the balance of trade? The frugal Mediterranean of tradition managed somehow with a herd of sheep and goats, some corn or barley, a little oil and a few fruits ... the rustic needs of the past. In exchange for a minimum of imports, the district would have quite enough early fruits and vegetables or citrus fruits to pay for its needs by its exports. But if it has to import more, then it must also export more.

That is exactly what is happening. As the years go by the needs of the Mediterraneans increase and become more complicated, because life keeps changing. The simple garden tools of the past are no longer enough, for intensive cultivation requires equipment that is becoming more and more specialised. Expensive heavy machinery which is difficult to manipulate must now be acquired. In this age of metallurgy and mechanics based on coal the Mediterranean does not make these things. In addition one must have all sorts of twentieth-century equipment such as motor cars, radios, and the thousand and one labour-saving devices which are being introduced everywhere.

The Mediterranean has become so completely wedded to this complementary system of trade that many of its districts are devoting themselves to a single crop like wine or flowers. The rare terraced wheat-fields that once were wrested from the mountainside have now become rarer still, and certain wine-producing districts even have no fodder for their animals. As all their products are sent off abroad, these districts recall the countries of the New World where one sees nothing but wheat, cotton or coffee, and where the farmer would be very much embarrassed if he had to depend on his own farm instead of living on the tinned goods bought in town on market day.

In these circumstances exports become a vital necessity. In the South American countries, where conditions are similar, people are even more dependent on foreign markets and are obliged to export on an enormous scale. But the Mediterranean by its very structure is not suited to mass production. It will never have the sensationally favourable trade balances of certain new countries where exports are frequently double the amount of their imports. How can the Mediterraneans ever hope to succeed since they are all doing the same thing, and are forced to compete keenly against one another for the attention of foreign buyers? Yet they simply must find markets for their wine, their oranges and lemons, and their oil.

They are thus thrown back necessarily on what the statisticians call 'services'. To pay for an article by trading it for another is not the only way, for one can also pay for it by rendering a service, though this is not quite so satisfactory. In the Mediterranean world we find deficits in the balance of trade made up in a dozen ways such as commissions, brokerage, transports and the funds that emigrants send back home to their families. In any case people are naturally ingenious in this part of the world. By time-honoured tradition the Mediterranean is an intermediary, a marvellously gifted agent who can act as a go-between, arrange everything, and find a solution when every avenue seems to be closed. The tourist trade also helps to right the adverse annual balance, since both winter and summer visitors bring in foreign exchange which is equivalent to exports. And besides, the very presence of foreigners provides countless opportunities for commissions and services of one kind and another.

The Mediterranean countries nearly always have a deficit in their trade balances, for over and above the articles which they do not manufacture they are obliged to import certain foodstuffs like cereals which are lacking, or raw materials like coal which are seriously needed. There are not enough exports to restore the balance, for they are neither available in quantity nor easy to dispose of when everyone else is competing for the same markets. So the Mediterraneans must have recourse to services as we have just said. To depend on tourists is always precarious, and even with their assistance it is difficult to right the balance.

It is hard to apply these observations to any particular country, for probably a large part of its resources will be derived from provinces on the continent which do not belong to the Mediterranean region either economically or in climate. This is true of France, Italy and Egypt, and to a great extent of French North Africa also. However, there is one country that is completely Mediterranean and that is Greece, so let us see what we can learn from her.

The Greek trade balance is chronically, indeed normally, unfavourable, for exports seldom go beyond 50 per cent of imports and usually do not even reach 50 per cent. The underlying reasons for this are obvious enough when we study the composition of this trade. Manufactures must be purchased abroad as there is little or no industry within the country. In the normal year of 1932, for example, manufactured articles amounted to 37.4 per cent of the imports; these consisted of cotton and woollen yarn and piece goods, machinery and spare parts, drugs and chemicals, perfumes, automobiles and other articles of current consumption without which anyone is at a loss, even people living as modestly as the Greeks. Yet the country is not self-supporting, for raw materials still account for 24 per cent in the trade balance, and there is always coal to be considered. Foodstuffs amount to 38.4 per cent, for cereals must be imported and also sugar. When these shipments stop famine soon follows, as we know to our cost.

To counterbalance these imports, which certainly are not luxuries, Greece has only her produce to dispose of, the same list as every other Mediterranean country: raisins, currants, wines, olives, nuts, figs, oranges and lemons and tobacco. These things cannot be produced in large quantities, and consequently trade can never right the balance entirely. But ingenious people like the Greeks always have other resources which can be classified as invisible exports: interest on capital invested abroad, profits from freight carried by the merchant marine, and other services which bring in commissions, brokerage and so forth ... These things are important in balancing the account, but still are not sufficient to maintain an equilibrium which is not provided for in the nature of things. Further support or financial aid must come from abroad.

Thus the Mediterranean economy is dependent on outside resources, in the sense that the soil does not yield enough and the people must rely on their own ingenuity and intelligence. The solution is not found by turning to the land, but rather by looking out to sea and to the economy of the world at large. This is suggested by the geographical structure, which presents a barrier inland but an open door towards the sea. The first natural association is between the coast and the interior, and it is on this combination that the political units of the Mediterranean are constituted. A little later, however, ties are bound with more distant lands and international economic relations are fostered. Thus coal is brought from England and wheat from Russia, the United States and Argentina; meanwhile it is towards northern Europe that the Spanish or Italian citrus fruits and the North African minerals are being exported.

Personally I cannot believe that any self-contained Mediterranean economy could ever exist. This sea is a great highway where commercial relationships are formed, and it is commerce not production that is the measure of its activity. It was not by chance even in ancient times that the hero of the Odyssey possessed all the qualities and some of the shortcomings necessary to successful trading. Whenever the deep roots of international commerce are severed, no economy suffers more than the Mediterranean. Though it is varied and resourceful,

THE BALANCE OF TRADE

it is not infinitely rich. In that it resembles its own charming climate, with its brilliant light and hot sun always with a chill in the air. If wisdom still has any value and there are still any philosophers left, one of them, as in ancient times, should draw a useful lesson from this idea.

CHAPTER XV

THE SANITARY DEFENCE OF THE WEST

THE geography of sanitation is an aspect of human geography which will be indispensable in the future, but nowhere more so than in the Mediterranean zone, that crossroads where the people of three continents pass to and fro.[1]

The principal maladies of the human race fall into three categories according to the way in which they are transmitted. Certain ailments are passed on directly from person to person, and consequently their area is the largest geographically as their diffusion is not limited to the existence of carriers. One can follow the way that tuberculosis, smallpox and syphilis have increased ever since the sixteenth century, and have travelled from continent to continent over the entire globe. These diseases are everywhere. Others like malaria or yellow fever, on the contrary, are linked up with certain carriers. They are most virulent in the tropics, and yet they also occur occasionally in sub-tropical and even in temperate countries. A third category includes ailments due to parasites like sleepy sickness (tripanozome) or bilharzia which are often carried by water. They are not usually found outside their own geographical domain where their particular parasite is able to prosper. The maladies in this class have a certain immovable quality.

The general distribution of infectious diseases throughout the world falls mainly into these divisions, but it is curious to note that their geographic distribution is zonal and horizontal, i.e. divided latitudinally. It is also interesting to find that their pathology, hygiene and medicine are zonal in the same way.

---

[1] Bibliography: *Destin des Maladies Infectieuses*, by CHARLES NICOLLE, and *Les Fondements Biologiques de la Géographie Humaine*, by MAXIMILIEN SORRE (Armand Colin).

THE SANITARY DEFENCE OF THE WEST

The tropical belt is the principal source of these maladies, but in spite of dangerous and offensive outbreaks the European actually does manage to protect himself against them. As for the sub-tropical zones, they have their own diseases which are due to their climate and social conditions, but in every way they are less virulent and their death rate is lower. They are an intermediate stage. And yet the sub-tropical zones are not free from infection that travels from the tropics by ships, caravans and even aeroplanes; the sub-tropical zones are in a way the front, the first to suffer. The pathological colour of regions like the Mediterranean is thus affected by epidemics which temporarily create a foreign medical atmosphere. Temperate and even cold countries have their own diseases, but these are chiefly associated with social conditions, and the rigours of a climate that is at once damp and chilly. The struggle is easier there and accordingly the diseases are less virulent.

Surveying the areas in which various maladies are inherent, M. Maximilien Sorre divides the globe into two basic regions: the Atlantic and the Pacific; and strange to relate he finds that up to the present little contact has ever existed between them.

He divides the Atlantic into a northern and a southern domain. The first coincides with the parts of the world which are densely peopled by the white race, with civilisations which are urban in character. The maladies which develop there are essentially of the 'social' type, such as tuberculosis, smallpox, syphilis and diphtheria; all these ailments are affected by meteorological factors and are liable to spread at an alarming rate. Geographically this domain consists of North America and Europe, in other words the very home of the white race. In this the Mediterranean is merely a sub-section of Europe. In the southern section of the Atlantic domain the tropical diseases are important, the African region being remarkable for the prevalence of sleepy sickness and the American for yellow fever.

As for the Pacific, it covers an island domain, a continental domain and a Euro-Asiatic domain. The island domain consists of Malaya and the Pacific Ocean itself and it is relatively healthy. But the continental domain on the other hand is the most formidable source of infection in the whole world. It is further subdivided into a Chinese and an Indian section. Without a doubt the Indian section presents the most propitious conditions for the maximum development of disease in all its complex forms — exceedingly dense populations huddled together in conditions liable to spread the worst possible contagion, and furthermore these people are usually undernourished and unaware of the first principles of hygiene. Cholera and plague are always present, and so are malaria and a dozen other diseases that are scarcely less serious. India seems to live under a curse. It is filled with wretched specimens of the human race, all hoping to be cured by a miracle. The Chinese sector also is a prey to epidemics, for there is the same lack of hygiene there. Bubonic plague rages, and leprosy spreads and ravages the Far East, and extends over great distances beyond that.

The Euro-Asiatic domain consists of an immense band of territory uniting northern Asia, the Amur region as far as the Don, the Black Sea and even the Red Sea. Ranging right across the steppes, diseases such as pneumonic plague, recurrent fever and leishmaniosis spread as far as the eastern Mediterranean, and even to India which certainly does not require these additional troubles.

II

The medical geography of the Mediterranean is due partly to the conditions prevailing in its vicinity, and partly to its own climate and social atmosphere.

The Mediterranean climate possesses certain striking advantages such as dryness and luminosity which are very stimulating. At the same time gentle breezes and kindly winds blow in from the sea, and there is always the purifying and invigorating effect of strong winds like the mistral. A climate of this kind demands little from the human organism. Consequently the

THE SANITARY DEFENCE OF THE WEST

Mediterranean does not require heavy meals and he is sober when it comes to alcohol. He is frugal in his living, though sometimes a little too close to the line of undernourishment. On the whole, however, he is remarkably well adapted to the efforts required from a man who is not industrial, and who needs ingenuity rather than strength in his work. We must not

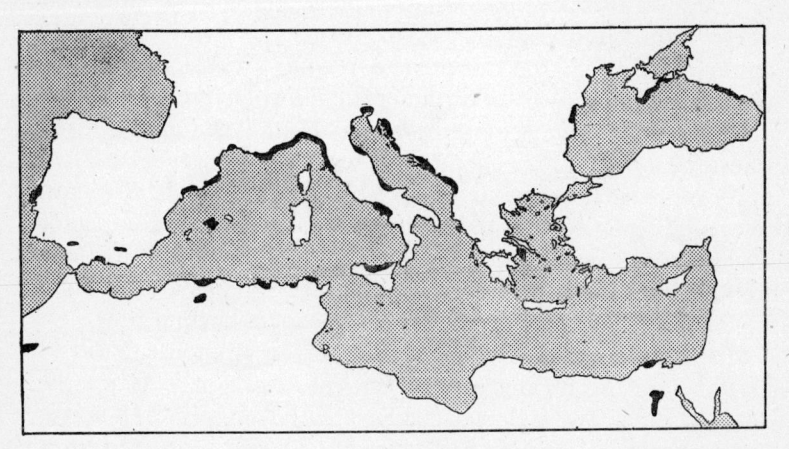

FIG. 35. WINTER RESORT AREAS ACCORDING TO MAXIMILIEN SORRE

overlook the fact that a climate that includes a succession of invigorating days when the mistral blows, and equally depressing days when the marin holds sway, is not likely to develop any energy. If the Mediterranean were faced with the intense industrial production of northern countries, no doubt his conditions of climate and nourishment would be insufficient to enable him to give a satisfactory economic output.

On the other hand visitors who come south to reside in the Mediterranean in search of better health find what they want without difficulty. This region has the finest winter resorts in the world, for in this respect the Côte d'Azur, the Italian Riviera, the coasts of Dalmatia, Naples, Sicily, Corsica, the Balearic Islands, Corfu, Algiers and Egypt cannot be surpassed

THE MEDITERRANEAN

(Fig. 35, p. 201). And one can easily add to the list, for the Mediterranean possesses such a magnificent asset in its resorts that the people of the northern countries have decided that even in summer it is hardly less attractive than in winter for those who are weary of rain, and longing for light, sun and warmth!

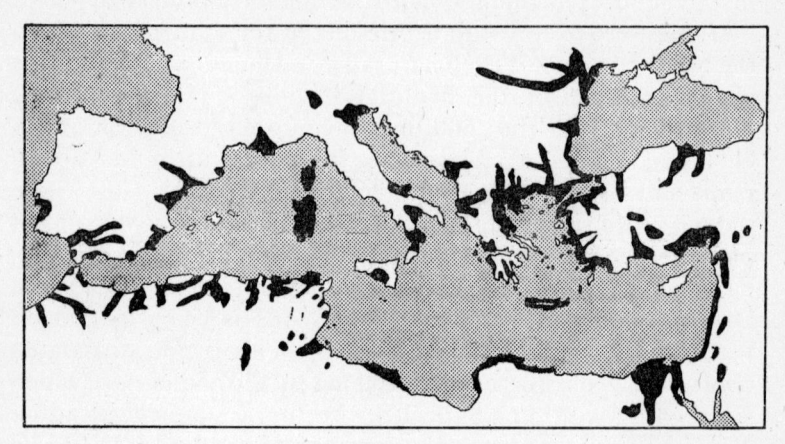

FIG. 36. ENDEMIC MALARIAL DISTRICTS, ACCORDING TO MAXIMILIEN SORRE

Visitors of course select their homes with care, and concentrate every effort on finding the exact hygienic conditions they require, but otherwise the Mediterranean is not uniformly excellent by any means. Apart from their newer suburbs the cities still have densely populated sections which are at the stage that M. Sorre courteously calls 'palaeotechnique', which simply means unhealthy. Indifference to hygiene is frequent, and it must be admitted that this is natural enough in countries where the sun and wind seem to be capable of dealing with any kind of infection. Yet this is rather an exaggeration, in fact in ancient times the townspeople were perhaps more aware of the drawbacks of a southern climate than we are. The resistance to epidemics today on the whole is pretty feeble. Malarial marshy districts are still the rule especially in the lowlands, and

THE SANITARY DEFENCE OF THE WEST

these are difficult to drain. As for foreign germs, they flourish owing to the contact with the East and with the tropics, both of which are close at hand ... The Mediterranean is the gateway to the Orient, the gateway to the desert, the gateway to the oceans, a meeting-place which is more exposed to contamination than Europe is. It is the first to be threatened, the first to be stricken, and therefore the first to be protected.

If one were to draw up a brief list of the ailments to which the Mediterranean is prey naturally we should start with those which originate in the malarial swamps (Fig. 36, p. 202). Unfortunately all the conditions favourable to malaria are to be found here and always have been — catastrophic downpours and floods which leave a residue of stagnant pools where the mosquito reigns supreme. All the three principal forms of malaria are present: benignant tierce, quartan fever and malignant tierce. However, the struggle against these diseases has begun, notably in Italy and France, and is being carried on with energy, persistence and success. This is not an isolated effort, but a constructive undertaking by the community which is linked up with the political regime and a stable civilisation. Nevertheless, even this conquest is precarious, and it would be jeopardised by a return to medieval conditions.

Among the maladies always threatening the Mediterranean one must first mention those which are associated with food, such as pellagra, a vitamin deficiency due to living on maize; or lathyrism, which paralyses the lower members and is due to eating too many peas. Both these ailments attack undernourished people especially. Then there are the various maladies which come up from the tropics, but which have become endemic in certain Mediterranean countries: dengue fever, elephantiasis and dysentery, which are chiefly prevalent in Egypt and Syria and generally as one approaches the Middle East.

The great epidemics originating in Asia are cholera, and bubonic plague which has its source in India. Yellow fever comes from central and southern America, but it penetrates by way of the West. These diseases have cropped up only on rare occasions in recent years, simply as episodes, but although they can be kept outside the limits of the Mediterranean region as a rule, they are always in the offing. Later we shall deal with the methods by which they are kept in check, and the administration which always must be on the alert.

Leprosy is still present, but today it flourishes only in Crete, Malta and on the Spanish coast. Trachoma and granular conjunctivitis now appear to be more menacing, especially in Egypt, southern Italy and the Greek islands where both have been increasing in virulence for some time. Recurrent fever or ague which is prevalent in Iraq, and bilharzia in Egypt are disabilities which are limited geographically to the countries of the Levant. Several others which occasionally become serious are spreading not only in the Mediterranean region but sometimes still further afield. These are pustule or Oriental sore, kala-azar (a disorder of the spleen), Maltese fever which is transmitted by goat's milk, exanthematic fever, and a feverish rash. Typhus, which is not confined specifically to the Mediterranean, continues to ravage these people because they are undernourished, debilitated and incapable of cleanliness.

As we readily conclude, most of these maladies come into the Mediterranean from elsewhere, and there they find conditions favourable to their development. Being in contact with Asia and the tropics, it possesses a foreign atmosphere medically, and so it is continually obliged to defend itself from contagion. It is threatened much more than Europe which is cold or temperate in climate. Problems of sanitary protection thus arise which apply particularly to the Mediterranean region.

III

Sanitary protection against foreign infections is in very fact essentially a Mediterranean problem, but it is also a European problem, in reality it is the defence of the Western world.

Asia is a fascinating continent. Its legendary wealth and its

THE SANITARY DEFENCE OF THE WEST

agglomerations of humanity haunt our imagination. But it is also a source of filth, corruption, contamination, and contagion which results in virulent epidemics. These epidemics would spread throughout the entire West if protection were not organised and rigidly maintained. Comparison with the Roman walls comes naturally to mind, the walls that protected the Empire against the raids of the barbarians. Today it is invasions of Asiatic infection which must be held in check. The comparison, however, is not strictly exact, for unlike Rome the West cannot set up a wall against the ailments of the East. It is more a matter of infiltration, in the course of which the Asiatic scourges are tracked down as close to their source as possible, closely supervised, and progressively checked by erecting barriers against them. The final gateway is the Suez Canal, which can be closed as a sanction as a result of the medical examination which is a matter of routine at this point. More often than not the suspected germs have been spotted and checked long before this. Suez thus remains the gateway to the East, the place where the West is left behind. There, according to Gobineau, one enters a world that is new and strange. Viewed from this angle, the canal is indeed a barrier like the walls of Rome. But protection against disease today is the result of a policy of preservation, much greater in portent than the simple wicket gate at Suez.

To understand how germs can be disseminated, one must also know the geographic routes along which Oriental diseases travel towards the West (Fig. 37, p. 206). The source of Chinese contamination lies at the beginning of one of the most important maritime routes of the globe, the itinerary running from Shanghai to Singapore, Colombo, Aden and Suez. The port of Singapore, the key to the Far East, is in another sense the key to the Indian Ocean and the navigation routes to the Red Sea, and thence towards the Mediterranean. The source of contamination in India spreads from Calcutta, Colombo and Bombay along the same highway and all in the same direction. It is therefore necessary to have exact information, above all

THE MEDITERRANEAN

rapid information, in order to warn the international sanitary authorities of the approach of suspected ships.

Another world highway that is also important for the spread of Indian diseases is that which goes through the Khyber Pass to Baluchistan, Iran and Mesopotamia. Caravans and isolated

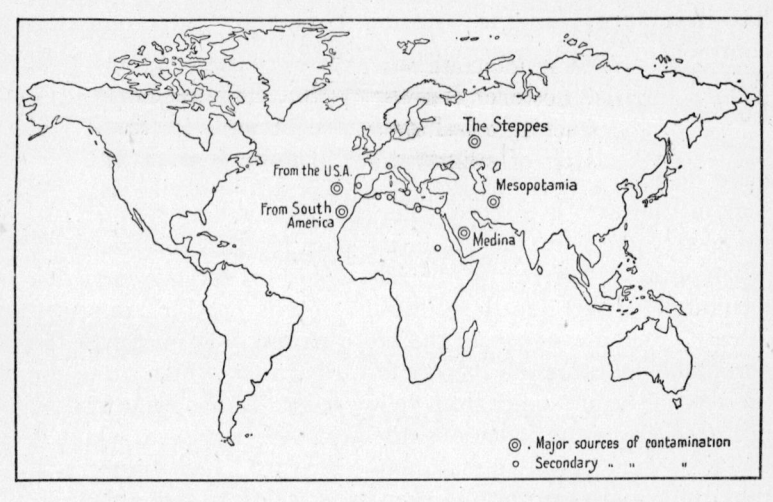

FIG. 37. SOURCES FROM WHICH DISEASES ARE INTRODUCED TO THE MEDITERRANEAN AREA, ACCORDING TO MAXIMILIEN SORRE

pedestrians follow this road and, through the half-ruined city of Meshed, wind their way to Kermanchah not far from Baghdad where the route divides into three. One branch then goes towards Istambul via the Baghdad railway, the second towards Damascus or Homs and the Syrian coast by way of the desert, and the third goes towards Basra and the Persian Gulf. From Basra, the contamination, travelling again by water, rounds the coast of Arabia, then along the Red Sea to end up at Suez.

There is still the threat of a third invasion. This time northern China is the home of infection which is propagated across the Tartar steppes in the direction of the Caspian Sea, thence to

THE SANITARY DEFENCE OF THE WEST

Russia towards the Black Sea, and then on towards the Mediterranean.

As for yellow fever which comes over the horizon from the other side of the world, for many years it had only one means of access, the navigation lines from central and southern America destined for Lisbon, Cadiz or Marseilles. Recently a new peril has developed, for the aviation companies have put the continent of South America in direct and rapid contact with West Africa, and also with Mediterranean Africa and southern Europe. The germs of yellow fever thus are transported so quickly that the former precautions based on the quarantine have become obsolete. Nor are Africa and Europe the only regions that are threatened. The air lines which on the eve of the war linked up the Atlantic coast of Africa either with Khartoum, or Egypt, or the east coast of Africa, opened a formidable outlet for the diffusion of this terrible American disease. The menace is all the more serious as from Egypt the aeroplane continues on its way towards India, which up to the present has been exempt from yellow fever. India certainly does not need any more infectious diseases! Air transport, which is bound to develop on an enormous scale in the near future, will thus oblige both Europe and even the Far East to redouble their precautions, and increase the severity of their medical supervision of the inter-continental highways.

We have indicated the great routes of access, but there are also on the edge of the Mediterranean zone, and indeed within the Mediterranean zone itself, secondary sources of germ dispersal. First of all there is Medina, with its pilgrims arriving from Mecca; then Benghazi in Tripolitania; and Kairouan in Tunisia. The great ports of destination, such as Barcelona, Marseilles or Genoa, are also affected occasionally, and from these points there is always the risk of disseminating germs still further afield. But a sound policy of preservation insists that the enemy should be checked much earlier than this.

As we have already explained, the visit of the doctor from the medical service at Suez or Port Said should be only the final

THE MEDITERRANEAN

verification. The real result of the present policy is the existence of a learned international organisation, whose network of information and precautions cover the face of the entire globe. The principal collecting points for medical information are at Singapore, Washington and Alexandria, and formerly the headquarters was at the League of Nations at Geneva. The International Office of Hygiene, which was established in 1907 by the Convention of Rome, is now charged with carrying out functions of the greatest importance. Its duty is to 'gather and bring to the attention of member states facts and documents of general interest to those concerned with public health, and especially infectious diseases, notably cholera, plague, yellow fever, exanthematic typhus and smallpox, as well as the measures being taken to combat these maladies' (Art. IV of the Statutes). 'The governments inform the Office of any measures they are undertaking to assure the application of the international sanitary conventions. The Office suggests modifications which might be advantageous to the working of the conventions' (Art. V).

The international medical service checking the spread of disease has hitherto been the care of a special institution, the Egyptian Maritime Council of Sanitation and Quarantine, set up by the Conference of Venice in 1892. Its personnel has been composed of international medical men, working in Egypt according to conditions laid down by the power in control of the canal. In other words, the international wicket gate to the West has been guarded by civil servants from the Western world. Since 1937 the capitulations have ceased to exist so far as Egypt is concerned, and as a result the Quarantine Council is giving place to a medical service which is entirely Egyptian in its personnel. The gateway to the East is no longer confided to Westerners.

Now the *limes*, the modern Roman walls, proved to be completely efficient under their international administration. The great epidemics that were the terror of the nineteenth century have been if not conquered at least held at bay. Plague,

THE SANITARY DEFENCE OF THE WEST

cholera and even typhus no longer are an immediate danger for Europe. The quarantines that were such a nuisance a generation ago are now rare. The Mediterranean is still exposed no doubt, but it is being well protected. To what must we attribute this magnificent success? The reasons are worth analysing. Medical science, in spite of its amazing progress, must not take all the credit. A great deal of merit is due to an international organisation, which is administered by the white race in conformity with the rules of accepted practice which have been conceived and carried into effect by Westerners. One almost wonders whether the actual administration in this case is not even more important than science!

Now the general direction of the world sanitation, which yesterday was still in the hands of Westerners, is today contested. The local powers wish to control this all-important wicket gate, if by the hazard of circumstances, it happens to lie within their boundaries. One does not question them medically for they are very competent. But must we, with our eyes shut, place in the hands of these newcomers to the Western world the task of this great administration on which depends in the last analysis the health of an entire continent? Seen from this rather special angle, the whole of the West is in jeopardy.

IV

In this study of sanitary problems, we have again touched upon some of the essential characteristics of the personality of the Mediterranean.

Once again the Mediterranean appears as a passage, bringing the West into contact with the East. Although typically European in many aspects, it is really only the threshold of Europe, situated as it is on the borders of two other continents. Owing to its mixed character, we find there, from the sanitary point of view, the best and the worst. The wonderful winter resorts are situated alongside malarial districts and are always in danger of foreign contagion. In the world of health, once again we are impressed with the Mediterranean's precarious position. The health resorts are simply the work of man, the result of civilisation, but even they are always in peril of returning to barbarism—and, moreover, the desert is not slow to claim its own. We recall Paul Valéry's saying which has already become a classic: 'As for our civilisations, we now know that we are only mortal.'

CHAPTER XVI

THE MEDITERRANEAN'S PLACE IN THE WORLD

THE Mediterranean is a route, but it is also a civilisation within which it has clung to its own conceptions across the years. As we come to the end of this study, let us try to give this long narrow sea its proper place among the great highways and ascertain its role in the equilibrium of the world. We shall also try to place it in relation to the various civilisations, and decide how far it is European, Oriental, or an integral part of the West. Geographically it is a crossroads, historically it is a source.

I

In the equilibrium of the world the function of the Mediterranean is to be an economic and political centre of gravity, either of the planet or of a network of great trade routes. This position has varied in different epochs. Let us therefore define the four main periods of history that have affected it: (1) ancient times up to the Middle Ages and Vasco da Gama; (2) the period of rivalry between the Mediterranean and the Atlantic; (3) the construction of the Suez Canal and the great days of the nineteenth century; (4) the effect of the canal and Europe in the twentieth century. Note that the point of view is never local.

At the time of the great Greek civilisation, the world's centre of gravity was situated in the eastern Mediterranean with a certain limited contact with Asia. Already the West was distinguishing itself from the East, and Greek culture in the sense in which we still understand it today was Western. Under Roman domination the Mediterranean became the centre of the Western world. The *mare nostrum* of the Romans was self-sufficient, self-supporting, and its contact with the outside world was simply a secondary aspect. The Arabian conquest divided this sea into two. Previously it had always been a single unit, but from then on there was both a Moslem and a Christian region. However, trade with the East continued. The Arabs meanwhile were developing the southern coast of the Mediterranean, and during the Middle Ages they brought more of the refinements of civilisation to the West. This was not a critical period for the Mediterranean; that came later and from another quarter.

In the second half of the fifteenth century two events of outstanding importance transformed the world: first, the Turkish conquest and the fall of Constantinople in 1453, after which the Mediterranean was no longer a passage but simply a blind alley. Secondly, at about this time a new maritime route to the Indian Ocean was discovered. The new route around the Cape of Good Hope was longer but it was safer, and moreover the journey was made without interruption. As a result in a few short years the Mediterranean was outstripped as an inter-continental economic route, and the centre of gravity of the world had shifted to the Atlantic coast. The Mediterranean then fell into a state of stagnation. The essential trade routes no longer went its way, and when the hour struck for the industrial revolution, the new and decisive techniques were not discovered on its shores.

After a long period of preparation, and as the result of political and commercial ideas which had been maturing in the Mediterranean since the close of the eighteenth century, the Suez Canal was opened in 1869, and the East and West were united by a new direct route. But trade had meanwhile changed completely. Rome, the Middle Ages and the West at the time of the Renaissance demanded from Asia – which they considered to be a land of miracles – luxuries, spices and precious stones of great value but of little weight. It was now a question – and it was this that assured the success of the canal – of an industrialised Europe, a Europe that had become an importer of raw materials to maintain her factories, and of agricultural products to feed her dense populations. Our equilibrium today still relies on this complementary trade, which is maintained by the regular shipping lines. Under this new regime the Mediterranean became, or rather had become once more, the quickest route between the countries of the West that had evolved economically, and the immense territory of the Far East that still has not been revolutionised industrially. Having at last rid itself of the piracy and barbarism which infested it for centuries, the Mediterranean became the essential artery of the world, and the vital axis of the British Empire. This period of triumph, however, lasted only from 1869 to 1914, scarcely half a century.

Now in the twentieth century Europe is losing its monopoly. Another world is arising outside its boundaries, and soon this new world will become autonomous, and will no longer revolve on the axis of the old continent. Its axis is the Panama Canal which was opened — historic date! — in 1914. This new canal is not competing against the Suez Canal which does not suffer in the least. It lies outside the orbit of Suez, and has its own relationships which have no bearing whatever on Europe.

Is it a simple coincidence that the Mediterranean should be returning once more to its old, its traditional lack of security? As a result of the war, and of new types of weapons such as the submarine and the aeroplane, it ceased to be safe from 1915 onwards, so much so that England re-routed her convoys via the old itinerary of Vasco da Gama. Britain now realises — and the war of 1939 confirmed this point of view — that a second route will be necessary, since the Mediterranean will be the arena of the great battles of the future. Renan's intuition surmised this as early as 1885, when in welcoming Ferdinand de Lesseps into the French Academy he said: 'Now that the canal has been cut Suez has become a strait, in other words a field of battle....' From that moment began what might be termed the crisis of the canal, the crisis of the Mediterranean, the crisis of Europe. This crisis moreover still is not over.

As a highway the Mediterranean is involved in this crisis, for the duel between Vasco da Gama and de Lesseps still goes on. (I should like to meet the spirits of these two great men, say in the Champs Elysées, and listen to what they have to say on the subject!) Knowledge of the future, even of the immediate future, is not revealed to us, but we can picture the conditions which are likely to take place. Now it is not an economic crisis, but a political crisis, which has compromised the route of the canal. If the day ever comes when international security is restored, Suez will naturally be the principal trade route between Europe and Asia. The Mediterranean will return to its proper place again, but that will not be while disorder and political disturbance run rife in this part of the world.

On the other hand, new technical possibilities suggest an immense development in the Mediterranean which is in no way connected with the Suez Canal. The motor car, the aeroplane and the railway are together forming a vertical axis which is bringing Africa and Europe closer. Conditions are today the same as during the Roman Empire when this sea, far from separating, actually united the various continents on its borders. In the same way trade between Europe and western Asia, since it has been rejuvenated by the new techniques of the motor bus, the aeroplane, and the pipe-line, can look forward to a magnificent development, and even extension into Syria, and on to the Persian Gulf and India. Finally, relations between Europe and America have ceased to depend entirely on maritime transport. No matter where the Clipper elects to come down, the trans-Atlantic air line cannot neglect the Mediterranean. From sheer necessity it will have to look for relays at Marseilles to serve central Europe and the Far East. At Algiers it will have to branch off towards central Africa; and it will need a station at Alexandria or Cairo from which to cover the Indian Ocean, Africa and the Far East.

Geological maps indicate the existence of the trough of the Mediterranean since the end of the Primary era. It has continued ever since, and still remains one of the most deeply engraved furrows in the structure of the globe. In its length ships will always be bound to sail through it. In its breadth, reduced to nothing by modern technique, the aeroplane flies over it as quickly as if it did not exist. This meeting-place of the winds as they swing round the four points of the compass conjures up possibilities for this sea that no historical circumstance can ever annul.

11

A civilisation exists in the Mediterranean which in its essentials is shared by all the countries along its shores. Everywhere one finds a certain conception of values, leading on to a certain conception of the individual and of the work done by the individual. Geography and climate seem to be the dominating factors, so that even after the rupture caused by the Arabian conquest this unity has still continued. As we have seen, the Mediterranean has an atmosphere of production which is distinctly its own, both geographically and chronologically. In situation it is Eurasian, bound by parentage with the old civilisations of Asia, in marked contrast to the new Anglo-Saxon communities overseas. When I have been in the United States I have often felt that the Mediterranean was at the opposite pole from North America. In the past, and even today in spite of its undeniable progress, the Mediterranean is at a pre-machine phase in which the tool is still predominant. Consequently this region is still in a sense archaic. 'A German', writes M. Jules Sion in the *Universal Geography* published by Armand Colin, 'finds it difficult to picture his early German ancestors, but a Mediterranean can see every day in his own harbours, and in his own fields, scenes which seem to be contemporaneous with Homer.'

The psychological characteristics which distinguish the Mediterranean are well known, but some of them are often exaggerated. Doubtless the *lazzarone* does exist in Naples, but he is not a typical citizen. There are few regions in the whole world where the farmer's effort is more intense or where work

THE MEDITERRANEAN

is accompanied by more prudence, thrift and frugality. Here we find no impressive collaborations carried out by vast numbers of people, but rather personal initiative, the ingenuity of the individual, and a genius for disentangling a baffling situation. By comparison the Anglo-Saxon, as soon as he is away from home and reduced to his own resources, seems to be awkward and ineffective. In his commercial dealings the Mediterranean shows a fertility of imagination and adaptation, in fact Ulysses of the *Odyssey* was the most accomplished example of a type which has always existed.

A born intermediary, an incomparable agent, the Greek is, today as yesterday, a most useful fellow in the East, capable of doing everything required, indispensable in fact. He unites in an astonishing way the qualities and even the shortcomings that are necessary to succeed. This suits the Mediterranean, for life in these parts would not be possible without the addition of commissions for endless services which allow him to make up for the deficiencies of the soil. In these surroundings, the Westerner from the point of view of intelligence usually seems to be at a disadvantage. He is less alert, and not nearly so quick as the Mediterranean who can calculate in a flash whether a certain transaction will pay him or not. It is often good business to be represented by a Greek or a Syrian, who will wriggle like a fish in water in the intricacies and intrigues of finance. But one should never entrust him with the general management of the affair! It will not be want of intelligence but rather too much. In the last analysis he fails because he lacks a sense of proportion, self-control and the ability to foresee and dominate the situation. In the long run it is this which distinguishes the Westerner, for even a mediocre Westerner is superior to a more gifted Oriental.

Although he belongs to the West the Mediterranean has never been properly fitted into the economic and political conditions of our day. He possesses an intelligent sense of association, and when his interests are at stake he knows how to make good use of his wits. One need only recall the blight that descended on his vineyards in Languedoc, and the remarkable way in which he extricated himself from his plight. The discipline of great organisations is not in his line. His groups and associations are limited, as if the geographical articulation of the countryside was reappearing in the social atmosphere. Family ties are strong and so is the clan, but otherwise political sense, although alert and well informed, often does not exceed the relatively narrow circle of the town or village.

Today there are no great states in the Mediterranean region: they are all inland beyond its borders and outside its centre of gravity. The Corsicans, the Berbers and the Greeks all carry on their rivalries with passion, but they are not constructive.

American theorists discussing the most recent and sensationally planned economy like to oppose the mass man of the future to the anarchic individual of a past which they reject as out of date. The theorist is quite certain that a shepherd from Pindus, a gardener from Provence, a little fisherman from the Italian coast and a Levantine tradesman ought all to be placed in the second category. To him they mean nothing more than their personal qualities mingled with their faults.

In the Middle East the word 'Greek' suggests a tradesman, and the mere term conjures up endless meanings, which again reflect a thousand facets of Mediterranean ingenuity. I have often been struck by the fact that according to statistics 72 per cent of the Greek people are listed as having no profession. Once one knows the Levantine one realises that 'no profession' stands for merchants, insidious and loquacious traders, innkeepers and café proprietors, cheap jewellers, travelling hairdressers, sweetmeat pedlars, traders in furs, dealers in watermelons, Parisian novelties and postcards, money changers, bankers, even camp followers. La Fontaine summed it up when he said: 'A Greek merchant in a certain country carried on his traffic. . . .'

But this is only one aspect of the Mediterranean. The winegrower of Languedoc now devotes himself to a single crop,

THE MEDITERRANEAN

and the big vineyards of Mitidja compare well with the immense enterprises of the United States. How are we to describe the great shipowners of Marseilles or Greece, or the cotton brokers of Alexandria? With amazing ease they keep abreast of everything that is happening. They are equally at home in London, New York and Shanghai. Ralli Brothers, another type of Mediterranean, hold their own with the most modern Americans. Here we must be on our guard, however, for perhaps the Mediterraneans are succeeding owing to those very qualities which man in the mass is liable to lose. Thus, in its own field, the Mediterranean economy produces the most singular incarnations. One sometimes doubts if some of these characteristics belong to the present day. Look at Greece: agriculture contemporary with Homer and then Ralli Brothers!

III

Should the Mediterranean be classified as part of the West? The essential traits of Western civilisation have often been analysed. We find first of all a conception of the knowledge and logical reasoning which has descended to us from the Greeks, and which distinguishes us definitely from the Orientals; then, a conception of the individual which also comes to us from the Greeks but from Christianity as well, and which the Renaissance and then the eighteenth century have admirably transposed in the social and political field; finally, there is the technique of industrial mass production which comes to us from north-western Europe, but which has only recently been made completely effective in the United States.

This Western civilisation has developed in a certain geographical atmosphere. Originally it arose in the Mediterranean. By contrast with the continental mass of Asia the life of Greece was maritime, distinctive and varied, and Rome which was more solid in structure was also within the Mediterranean. It was later on that a civilisation was formed in north-western Europe which was at once new and yet traditional, and it is according to this civilisation that the West lives today. From it came the industrial revolution, as well as the immense mechanical development of modern times. Even if the home of technical progress tends to centre itself in the New World, it is in the Old World that we must look for its first impulses. Certain geographical circumstances are thus easily traced to the origins of the West and in this the Mediterranean has played an outstanding role.

In the first place I see in Europe, but essentially in the Mediterranean, a geographical setting which is of the measure of man, where Nature is neither crushing and overpowering, nor out of proportion to mankind. In such an environment the human being bears a relation to the land in which he lives, and Europe is the only continent of which this can be said. The consequences are obvious.

Although all white people are not associated in this, for those in Asia have remained aloof, Western civilisation is the work of a single race. It is the work of the white race, and it is they alone who have created the West. The distance which separates them from the blacks and the reds is immense, and although the yellow race may be capable of comparable efficiency they are still far behind in the technical field. The Japanese are the only people who have caught up, and then only in certain productions, with the handicap the white race has gained during the past three centuries. In any case there are certain qualities without which the progress of the West would have been impossible, and these qualities are possessed only by the white race.

From these remarks one comes to the conclusion that the geographical domain of Western civilisation should be defined precisely. In ancient times the East could already be distinguished from a region which still was not called the West but one to which Greece was later to give a new intellectual inspiration. When the 'free men' of Greece resisted the Persians they outlined the frontier which history has handed down to us. The Aegean Sea and the Greek cities of Asia corresponded to a civilisation which was distinct from the Persian, for the latter belonged to the Orient.

When Hellenism under Alexander was extended as far as India, it looked as if the frontier had been pushed 1500 miles towards the East. This was only a temporary tide, however, for at the close of the Roman Empire the East reclaimed its rights, and even Orientalised Hellenism itself. In the second century A.D., the *limes* of Rome were set up in Syria along the edge of the desert, for a certain geographical limit always tended to reappear. With the Arabs it was the East in its turn which overflowed the West right across the Mediterranean as far as the Atlantic Ocean. The Arabs proved to be more civilised than the Europeans, for it was they who possessed the qualities of initiative and intellectual liberty, thanks to which the West later raised itself to its outstanding position. Finally with the decline of the Ottoman Empire, the boundaries of the past reappeared, in fact they were not very different from those in existence today.

In the eastern Mediterranean it is still possible to trace the final frontier of the West. As a matter of fact the line will be that which divides the Mediterranean zone from the interior of the continent. The Mediterranean, no matter how much it may be influenced by the East, will always be partially dependent upon the West. On the coast the ports which are known as 'échelles' or seaports belong to the West, as opposed to the entrepôt markets of the interior. The latter are really the ports of the desert and are usually described as bazaars. While Alexandria, Beirut and Tripoli are Mediterranean in character, Cairo, Jerusalem, Damascus and Aleppo live under another climate, have another atmosphere, and belong to another world. We have here a contrast between two geographical zones, between two civilisations. The Mediterranean has well been described as the anti-desert. Let us add that the East begins with Islam, which has taken back all that Greece had conquered in Asia. In point of fact, the life of the desert, of those great arid spaces of the earth, can never be part of the

West. They belong to the East, and therefore they must return to the East.

Thus the Mediterranean stands in direct opposition to the Orient, even where it is under eastern influence. But, on the other hand, the Mediterranean civilisation, derived from the Greek tradition, is also opposed to the civilisations based on mass production and mass thinking, which are now at the head of the modern movement of the West. The latter are founded either on the automatism of the machine, which they substitute for the initiative of the individual; or on the collective action of widespread groups, instead of personal action or the flexible co-operation of restricted groups; or on the standardisation of the factory lathe rather than on the fantasy of the tool.

The countries which lend themselves to mass production are those with unlimited natural resources, of massive structure, and where vast areas of land extend in unbroken stretches. They have powerful national organisations, and their great markets are not divided up into compartments like ours. Regions like the Mediterranean are simply not in the running at a time when the world is turning more and more to the massive continents. Mediterranean workers are craftsmen rather than labourers; its cultivators are gardeners rather than peasants; its genius lies in commerce rather than in heavy industry; here weight, pure and simple, makes no appeal. Owing to its very qualities it is inclined to resist an evolution which seems to be too swift. Many of its institutions still bear the mark of ancient traditions, in which it is often possible to discern survivals of Arabian or Roman times. The idea of the clan, of the head of the family, of patrimony, of the solidarity of the family, the Mediterranean conception of the special role of the woman — all these things belong to the past rather than to the future. There are advantages in this archaism, for it maintains the prestige of certain values, values which the Westerners need, and to lose which would mean nothing short of ruin.